BUILDING YOUR OWN DRONES

A Beginner's Guide to Drones, UAVs, and ROVs

John Baichtal

800 East 96th Street,
Indianapolis, Indiana 46240 USA

Building Your Own Drones: A Beginners' Guide to Drones, UAVs, and ROVs

Copyright © 2016 by Que Publishing

ISBN-13: 978-0-7897-5598-8
ISBN-10: 0-7897-5598-X

Library of Congress Control Number:

Printed in the United States of America

First Printing: August 2015

Trademarks

All terms mentioned in this book that are known to be trademarks or service marks have been appropriately capitalized. Que Publishing cannot attest to the accuracy of this information. Use of a term in this book should not be regarded as affecting the validity of any trademark or service mark.

Warning and Disclaimer

Every effort has been made to make this book as complete and as accurate as possible, but no warranty or fitness is implied. The information provided is on an "as is" basis. The author and the publisher shall have neither liability nor responsibility to any person or entity with respect to any loss or damages arising from the information contained in this book.

Bulk Sales

Que Publishing offers excellent discounts on this book when ordered in quantity for bulk purchases or special sales. For more information, please contact

U.S. Corporate and Government Sales
1-800-382-3419
corpsales@pearsontechgroup.com

For sales outside of the U.S., please contact

International Sales
international@pearsoned.com

Editor-in-Chief
Greg Wiegand

Executive Editor
Rick Kughen

Development Editor
Ginny Bess Munroe,
Deadline Driven
Publishing

Managing Editor
Kristy Hart

Project Editor
Elaine Wiley

Copy Editor
Bart Reed

Indexer
Erika Millen

Proofreader
Jess DeGabriele

Technical Editor
James Floyd Kelly

Publishing Coordinator
Kristen Watterson

Interior Designer
Mark Shirar

Book Designer
Mark Shirar

Compositor
Nonie Ratcliff

Contents at a Glance

Table of Contents

About the Author

John Baichtal has written or edited over a dozen books, including the award-winning *Cult of LEGO* (No Starch Press, 2011), *LEGO hacker bible Make: LEGO and Arduino Projects* (Maker Media, 2012) with Adam Wolf and Matthew Beckler, *Robot Builder* (Que, 2014), and *Basic Robot Building with LEGO Mindstorms NXT 2.0* (Que, 2012). His most recent book is *Maker Pro* (Maker Media, 2014), a collection of essays and interviews describing life as a professional maker. John lives in Minneapolis with his wife and three children.

Dedication

This book is dedicated to my Grandma Marion, who is a couple months shy of her 98th birthday as I write this. She was in the hospital a few weeks ago with heart problems and the doctors told her to get her affairs in order, and sent her home in hospice. Grandma isn't ready to leave the party, however, and she's been feeling better, buoyed by the great care she's received from my mom and aunt. Her love of life and passion for writing give me strength every day.

Acknowledgments

When thinking of my family, I am confronted by two irrefutable facts:

1) Arden, Rosemary, and Jack are the best kids anyone could ask for.

2) None of this would mean anything without my dear Elise. I love you!

Workwise, thanks for the inspiration and assistance to Windell H. Oskay, Johngineer, Matthew Beckler, Riley Harrison, David Lang, Trammell Hudson, AnnMarie Thomas, Pete Prodoehl, Bruce Shapiro, Alex Allmont, John Edgar Park, Dexter Industries, Miguel Valenzuela, Pete McKenna, Steve Norris, Steven Anderson, MakerBeam, Jude Dornisch, SparkFun Engineering, Brooklyn Aerodrome, Adam Wolf, Michael Freiert, Sophi Kravitz, Christina Zhang, Lenore Edman, Rick Kughen, Sean Michael Ragan, John Wilson, Susan Solarz, Akiba, Mark Frauenfelder, Chris Berger, Michael Krumpus, Alex Dyba, Brian Jepson, Becca Steffen, Dave Bryan, Actobotics, Mike Hord, Makeblock, Pat Arneson, and Erin Kennedy. Apologies to anyone I forgot!

My mom, Barbara, compiled the Glossary, and I am forever indebted to her for helping out, in this and so many other things.

We Want to Hear from You!

As the reader of this book, you are our most important critic and commentator. We value your opinion and want to know what we're doing right, what we could do better, what areas you'd like to see us publish in, and any other words of wisdom you're willing to pass our way.

We welcome your comments. You can email or write to let us know what you did or didn't like about this book—as well as what we can do to make our books better.

Please note that we cannot help you with technical problems related to the topic of this book.

When you write, please be sure to include this book's title and author as well as your name and email address. We will carefully review your comments and share them with the author and editors who worked on the book.

Email: feedback@quepublishing.com

Mail: Que Publishing
         ATTN: Reader Feedback
         800 East 96th Street
         Indianapolis, IN 46240 USA

Reader Services

Visit our website and register this book at quepublishing.com/register for convenient access to any updates, downloads, or errata that might be available for this book.

Introduction

Drones are in the news all the time—and let's face it, they're likely to be an increasing part of our lives. We can throw on a tinfoil hat and look for small helicopter-shaped shadows, or we can learn as much as we can about these interesting devices. I suggest the latter—there is a lot of cool technology out there, and the best way to control it is to understand it.

Who This Book Is For

Aspiring drone-builders of all stripes will appreciate this book, as it covers many different areas of building your own drone projects, including not only electronics, but motors, airframe-building techniques, and tools.

How This Book Is Organized

This book consists of a main project, a quadcopter you'll build over the various chapters. The alternating chapters describe a variety of projects such as a data-gathering rocket drone, a blimp, and a boat made out of soda bottles, giving you a perspective on drones beyond those quadcopters that have everyone abuzz.

- Chapter 1, "A History of Drones," consists of a history of drones and brings you up to speed on current technological limits and terminology drone pilots use.
- Chapter 2, "Showcase of Cool DIY Drones," describes a dozen cool drones, including UAVs (unmanned aerial vehicles), ROVs (remotely-operated underwater vehicles), and rovers built by hobbyists and amateurs alike.
- Chapter 3, "Overview of Commercial Drones and Kits," introduces a number of commercial drones that you might care to purchase. Everything from a quadcopter packing a video camera to an undersea explorer is on the table.
- Chapter 4, "Building a Quadcopter I: Choosing an Airframe," begins the quadcopter project as you learn about a variety of airframes and chassis products, and you begin building your quadcopter's airframe out of a kit.
- Chapter 5, "Rocket Drone Project," breaks from the quadcopter and has you build a rocket drone, a model rocket with a basic Arduino payload.

- Chapter 6, "Building a Quadcopter II: Motors and Props," discusses two key components of your quadcopter build. You are presented the various options for purchasing motors and propellers, and you are shown how to mount them onto your quadcopter's airframe.
- Chapter 7, "Blimp Drone Project," shows you how to build a blimp drone, a small wooden robot hoisted aloft by helium balloons.
- Chapter 8, "Building a Quadcopter III: Flight Control," shows you how to control your robot while its in the air, with flight controllers and electronic speed controllers doing most of the work.
- Chapter 9, "Drone Builder's Workbench," covers the various tools I used to build the projects in the book.
- Chapter 10, "Building a Quadcopter IV: Power Systems," introduces a very important topic: how to power your quadcopter. This includes instructions on building a power distribution system to deliver electricity to the motors.
- Chapter 11, "Waterborne Drone Project," demonstrates how to make a simple remotely-operated vehicle build out of soda bottles.
- Chapter 12, "Building a Quadcopter V: Accessories," covers the variety of accessories, such as camera mounts, available for purchase or creation.
- Chapter 13, "Making a Rover," shows you how to make a rolling robot that uses RFID tags to navigate.
- Chapter 14, "Building a Quadcopter VI: Software," profiles some flight control software and autopilot firmware and also explores the ins and outs of the control software of the autopilot we used in the copter project. With the conclusion of the book, you will complete the quadcopter build.
- Finally, the Glossary explains the various terms used throughout the chapters.

If you have any questions, or want to learn more about the projects and my other books, the best way is to check out my Facebook page, www.facebook.com/baichtal. You can also email me at nerdyjb@gmail.com or follow my Twitter feed @johnbaichtal. Good luck and happy drone building!

A History of Drones

Imagine a car without a driver or a plane without a pilot, with a computer replacing the person who operates the vehicle. You're imagining a drone.

Drones are everywhere in the news, especially with battlefield stories about unmanned aerial vehicles (UAVs) shooting missiles at targets half a world away from the person pressing the button. However, not all drones are used in war. Some drones are peaceful drones.

Budget cuts at NASA have thrust these remotely operated probes into the spotlight—first and foremost the Mars rovers (see Figure 1.1). These remotely controlled rollers have performed vastly beyond the expectations of NASA's engineers.

FIGURE 1.1 An artist's concept drawing of a Mars Rover (credit: NASA/JPL/Cornell University).

Governmental use of drones is one thing, but do amateurs also use drones? The answer is yes. Ordinary hobbyists, tinkerers, and small business owners make and operate their own drones. For example, winemakers fly camera-equipped quadcopters (miniature four-bladed helicopters) to observe the state of their foliage in the arbors, without leaving their house. Other entrepreneurs flying similar quadcopters have upended the aerial photography business by eliminating the need for full-sized helicopters. Amazon and other companies are exploring package delivery by drone.

The purpose of this book is to introduce you to the current state of quadcopters, UAVs, ROVs (remotely operated vehicles), and other variants of the technology, with an opportunity to work on simple drone projects such as a rocket with an accelerometer aboard it, a water-borne drone, and a blimp bot made from Mylar balloons. Simultaneously, you'll follow along as I detail the careful assembly of a quadcopter, thus enabling you to build your own or learn how to make one even better.

What Is a Drone?

Let's get one thing clear: The definition of "drone" isn't really all that clear.

Drones get their name from honeybee drones, which go about their tasks mindlessly, as they are controlled by a faraway queen. Similarly, a robotic plane with a microcontroller programmed to work as an autopilot works much the same way, albeit with the help of technology.

Devices described as drones fall into two basic camps. First, there are *autonomous robots* whose operators take active control as needed. The rest of the time, autopilots take the lead, theoretically allowing a single operator to manage multiple craft. However, when the need arises, the operator can disable the autopilot and take back control.

The second camp involves quadcopters and other "helicopter-esque" flyers. The public sometimes calls these drones, despite the fact that most of them are merely radio-controlled (RC) models and not autonomous. Perhaps the reason why the two definitions have merged is that multirotors have recently become a great platform for autopilot-driven, microcontroller-based autonomous flight.

Hobbyists are operating swarms of quadcopters, creating new games where the drones compete against each other, and they are packing everything from cameras to barometric sensors and ultrasonic rangefinders into their creations. Meanwhile, a huge education market has developed, with teenagers and younger children building autonomous robots using building sets such as LEGO Mindstorms and VEX.

We're in the infancy of a cool phenomenon, and we can be a part of it! Let's build some drones together.

Three Terrrains

Drones are differentiated by the terrain the vehicle traverses:

■ Unmanned aerial vehicles (UAVs)
■ Remotely operated vehicles (ROVs)
■ Rovers

These three types of drones are discussed in the following sections.

Unmanned Aerial Vehicles

The term *unmanned aerial vehicle* describes drone airplanes (such as the Predator shown in Figure 1.2) and helicopters. Pretty much, if it operates in the air, we'll call it an UAV. The most popular hobbyist UAV is the quadrotor or quadcopter, and because of its popularity, this book will focus on this type of UAV.

FIGURE 1.2 The Predator drone did much to teach the public about drones and how they work (credit: U.S. Air Force).

UAVs most often are controlled through radio waves, such as the signals generated by an RC handset. Others use Wi-Fi or cellular technology to communicate. Many also include GPS receivers so their flights can be tracked on a map.

Remotely Operated Vehicles

A remotely operated vehicle is an underwater drone, usually tethered to a boat or submersible by a data wire, which is necessary because radio waves are dramatically hindered by water. ROVs have been used by ocean explorers for years. You can see an example of an ROV in Figure 1.3.

FIGURE 1.3 The OpenROV explores a shipwreck (credit: OpenROV).

Rovers

A *rover* is an RC car with extra features. It rolls around and navigates an earthbound terrain using sensors to detect obstructions. Rovers often feature tank treads or knobby tires, like the one shown in Figure 1.4. This helps them traverse uneven ground. Being ground-based gives rovers the capability to use all sorts of sensors to navigate, including ultrasonic, RFID, and bump sensors. You'll build a rover in Chapter 13, "Making a Rover."

FIGURE 1.4 This camera-mounted rover sports knobby tires for rough terrain (credit: Geoffrey Irons).

Anatomy of a Drone

Every homemade drone will differ from the next; that said, most drones have a number of features in common. The following are commonplace quadcopter components. Follow along with Figure 1.5 to see how each part fits into the full project.

A. **Props**—The props of a quadcopter typically consist of two standard and two "pusher" props rotating in opposite directions.

B. **Motors**—Quadcopters use DC or AC motors. There are countless varieties and price points, with premium motors catering to wealthy tinkerers. In Chapter 6, "Building a Quadcopter II: Motors and Props," I go into detail on a number of great hobbyist-friendly motors.

C. **Electronic speed controllers (ESCs)**—ESCs convert DC to AC for brushless motors, and also trigger the motors' power supply. You'll need one for each motor. ESCs' firmware can be modified to create different motor behaviors. For instance, ESCs are often configured to slow down the motor rather than stopping abruptly.

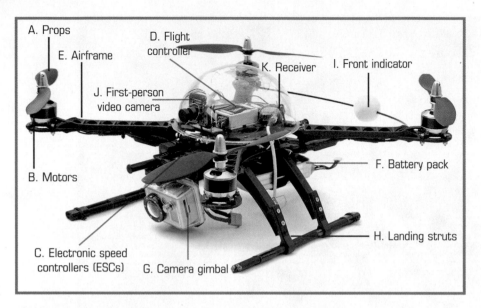

FIGURE 1.5 The quadcopter has a lot of parts and equipment (credit: Steve Lodefink).

D. **Flight controller**—The flight controller assists manual flight with certain autonomous functions. For instance, many flight controllers have a tilt sensor that keeps the drone level. Oftentimes flight controllers will have a certain pre-programmed routine they perform if the quadcopter leaves the control range.

E. **Airframe**—*Airframe* is the proper name for the drone's chassis. The airframe consists of a number of different elements, including motor booms as well as an enclosure or platform for housing the electronics.

F. **Battery pack**—Often a LiPoly battery, the robot's battery pack keeps the propellers turning while also powering whatever electronics are onboard.

G. **Camera gimbal**—This is a rotating platform on which a video camera is mounted. Servomotors allow the operator to turn and angle the camera during flight.

H. **Landing struts**—Quadcopters with a camera gimbal or other protuberance on the bottom need landing struts, which are little legs the drone rests on when it's on the ground. On the other hand, drones without gimbals often don't need struts, and simply land with the entire airframe on the ground.

I. **Front indicator**—Quadcopter operators need to know the front of the hovering aircraft, and it might not necessarily be obvious. Multiple solutions are available, ranging from differently colored props, LEDs, and reflective material—or in the case of the quadcopter in Figure 1.5, a colored ball that marks the *rear* of the craft. All that matters is that it makes sense for you!

J. First-person video camera—Low-resolution camera that sends its images to a ground station via radio waves.

K. Receiver—This small box translates radio signals into instructions for the flight controller.

Summary

In this chapter, you learned about drones, their common configurations, and the various components that go into them. In Chapter 2, "Showcase of Cool DIY Drones," you'll see what other folks have done with the technology. You'll be surprised at the diversity of all of the cool projects out there!

Showcase of Cool DIY Drones

So you want to build your own drones? Awesome. The best way to start is to admire the work other people have done. The following 12 projects are just a sampling of all the cool DIY (do-it-yourself) work you can find out in the wild!

Bicycle Rim Quadcopter

This project bears out the truism that pretty much anything can be made into a quadcopter's chassis, so long as it's reasonably light and strong. Built by Sam Ley, this quadcopter (pictured in Figure 2.1) flies very well and has survived several crashes.

You'll buy or build your own airframe in Chapter 4, "Building a Quadcopter I: Choosing an Airframe," but as you make your decision, look back on this crazy quadcopter and know that you have a lot of options.

3D-Printed Mini Quadcopter

This airframe, designed by Thingiverse user Brendan22, comes in a variety of configurations, including the four-bladed mini-drone shown in Figure 2.2. Another, the T-6, consists of three booms with two motors and propellers each. You can find Brendan22's designs at the following website: http://www.thingiverse.com/Brendan22/.

This is another example of a DIY airframe that you should consider when you make your own in Chapter 4. You can save yourself tons of time by taking advantage of free resources like Thingiverse to simply manufacture the parts you want. Simple if you have a 3D printer, of course!

FIGURE 2.1 Sam Ley made clever use of found material to make his quadcopter's airframe (credit: Sam Ley [CC-A]).

FIGURE 2.2 Have an idea for an airframe design? Just print it out! (credit: Brendan22)

Clothesline Racer

This autonomous robot, built by Mike Hord, follows a clothesline or cable to its terminus, then reverses course. It's a very simple drone, but a drone nonetheless! It has a very rudimentary control system—an ultrasonic sensor that tells the microcontroller to reverse the motor—and no steering. You can see the racer in Figure 2.3.

FIGURE 2.3 This autonomous line-following robot goes back and forth (credit: Pat Arneson).

The racer goes to show that drones can look like anything. Don't be roped into just one category when building a drone. Four chapters of this book detail non-quadcopter drones, including a rocket, a blimp, a boat, and a car.

Vessels

Vessels is a project by Stephen Kelly, Sofian Audry, and Samuel St. Aubin, consisting of dozens of small floating robots (shown in Figure 2.4) that scoot around a pool, communicating with each other with infrared signals and audio tones. The idea is that they're almost behaving like emergent living beings. You can learn more about the Vessels project at http://vessels.perte-de-signal.org/project/.

Chapter 11, "Waterborne Drone Project," details a floating drone that uses a computer fan to move around. Simple, slow-moving boats like these are great for backyard experimentation, because you can test it out very easily!

FIGURE 2.4 These autonomous robots actually act somewhat like living beings (credit: Beatriz Orviz, LABoral [Spain]).

Radio-Controlled Blimp

This blimp, built by robotics students and instructors at Idaho State University, uses two DC motors for the propellers, each angled by a servomotor so it can turn independently (see Figure 2.5). The operator controls the blimp using a custom handheld control unit; XBee wireless cards in the blimp gondola and controller communicate with each other. You can learn more about this project at http://www.thingiverse.com/thing:98815.

In Chapter 7, "Blimp Drone Project," you'll have your own chance to build a blimp, using a laser-cut wooden gondola supporting a pair of motors, with a radio-control receiver on-board.

FPV Quadcopter

Steve Lodefink's beautiful copter never looked this good again—mostly because the next flight after this photo was taken, it suffered a failure, fell out of the sky, and smashed to pieces on the ground (see Figure 2.6). It had two cameras: a low-resolution FPV (first-person video) camera that transmitted a picture over the radio waves, and a GoPro Hero2 camera for high-resolution shots.

A beautiful drone like this one demonstrates why it is such a popular category. You'll build a quadcopter (smaller and simpler than Steve's) in stages throughout this book.

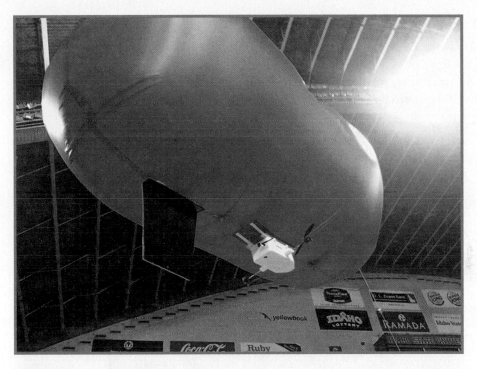

FIGURE 2.5 This blimp features a 3D-printed cabin (credit: Geran Call).

FIGURE 2.6 This lovely quadcopter lost power and dropped to the ground (credit: Steve Lodefink).

Open RC Trike

The trike shown in Figure 2.7 has a 3D-printed body and wheels, with a servo controlling the steerable front wheels and a motorized rear wheel pushing the vehicle along. A classic RC rig controls the steering and movement. You can find the design files at http://www.thingiverse.com/thing:499130.

FIGURE 2.7   This RC trike has a 3D-printed chassis (credit: cupidmoon).

Drones that roll on the ground are called rovers, and you'll build one in Chapter 13, "Making a Rover." Rovers have a completely different set of priorities and challenges than flying robots, and are a lot of fun to build.

Foldable Quadcopter

Roger Mueller designed and printed his quadcopter's airframe to make it easy to fold up so he could take it on hikes. In Figure 2.8, you can see the quad after it had fallen 20 meters— the only thing that happened was the landing struts broke off. You can find the design on Thingiverse.com, http://www.thingiverse.com/thing:71972.

One thing you should take into consideration when designing and building drones is that they crash! Every bird ever built has fallen out of the sky at least once. Chapter 12, "Building a Quadcopter V: Accessories," details some add-ons such as a parachute and a plastic gondola that help protect a drone from impact.

FIGURE 2.8 Roger Mueller's foldable quadcopter survived a fall that broke its landing struts (credit: Roger Mueller).

Mini-Quadcopter

Steve Doll's "SK!TR" quadcopter, shown in Figure 2.9, is as big as your palm (not counting the motor booms), and it uses an open-source flight stabilizer called OpenPilot CopterControl (openpilot.org). Steve runs his own quadcopter store (hovership.com) where he sells motors, airframes, and complete kits.

Chapter 8, "Building a Quadcopter III: Flight Control," details a number of other autopilots, also known as flight controllers. You'll add a commonplace controller to your own quadcopter.

FIGURE 2.9 Steve Doll's SK!TR quadcopter is small enough to fit inside a lunchbox (credit: Steve Doll).

3D-Printed RC Boat

Michael Christou's 3D-printed boats race through the water with propellers, impellers, and other modes of propulsion. A former engineer now in retirement, Michael gets his creativity out by designing cool-looking vehicles like the boat shown in Figure 2.10. To learn more, go to http://www.thingiverse.com/thing:272132.

I already mentioned Chapter 11, where I talk about waterborne drones. In that chapter you'll learn about a very important part of building boat drones: waterproofing them.

FIGURE 2.10 Michael Christou's RC boat designs can be downloaded and printed (credit: Michael Christou).

Tricopter

The carbon-fiber UAV shown in Figure 2.11 was built around a MultiWii flight controller, a very inexpensive open-source board. Tricopters work differently than quadcopters because one of the motors isn't fixed. The rear propeller of the tricopter moves with the help of a servo, allowing for extremely maneuverable flight. In addition, the tricopter has a mount for a GoPro video camera on the front. To learn more, check out http://theboredengineers. com/2013/07/the-tricopter/.

This tricopter project illustrates a phenomenon of the multicopter category: There are many interesting configurations besides the "regular old" quadcopter to explore, including six, eight, and yes, three props.

FIGURE 2.11 This tricopter is highly maneuverable thanks to its moveable third motor (credit: theboredengineers).

Mecanum Wheel Rover

Wheels are important for rovers, because that's how they get around! The 3D-printed rover shown in Figure 2.12 ups the ante by sporting mecanum wheels, which consist of motorized wheels with smaller, unpowered wheels around the edge. These wheels allow for movement in any direction. The rover is controlled by a Chumby One microcontroller. You can learn more at http://www.thingiverse.com/thing:5681.

You'll learn about a variety of wheels and trade in Chapter 13. The topic is just as important to rover builders as choosing a propeller is to a quadcopter flyer. The good news is that there are a lot of cool options to buy or make.

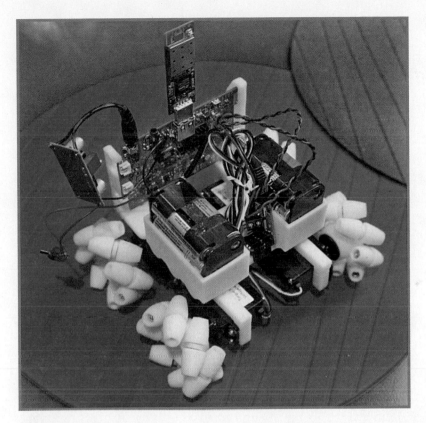

FIGURE 2.12 This rover uses mecanum wheels to go in any direction (credit: Madox).

Summary

Hopefully this chapter has given you a sense of the cool projects floating around out there—literally, in some cases. You had a chance to learn more about a variety of drones, including quadcopters, boats, planes, and cars. In Chapter 3, "Overview of Commercial Drones and Kits," you'll get to see some kits and finished drones that are a lot like the ones in this chapter, but considerably more polished.

Overview of Commercial Drones and Kits

The easiest way to get a drone, as usual, is to buy one. There's nothing wrong with buying a pre-made product because it gets you flying faster. However, you don't really learn anything about electronics or robot building. A kit, by contrast, gives you the opportunity to plug in the various components. In doing so, you learn a little about the parts and how they fit together. It's a nice middle ground between doing everything yourself and just buying a finished bird.

Parallax ELEV-8 Quadcopter

The ELEV-8 (P/N MKPX23) is the second version of Parallax's entry into the world of quadcopters. Parallax makes microcontrollers, most noticeably the Propeller, which features the PBX32A chip. Not coincidentally, the flight controller for the ELEV-8 also features the same chip. The quadcopter, therefore, is something of a plug for Parallax's hardware.

The ELEV-8 is cleanly designed (you can see this in Figure 3.1), and it conceals the usual ugly wires and zip ties by hiding them inside the hollow booms, while the ESCs are tucked between the two central mounting plates (see Figure 3.2).

Be aware that Parallax doesn't include a battery pack, an RC transmitter, or an RC receiver. Obviously you'll need to buy your own in order to fly. The company suggests a specific lithium poly-mer (LiPo) battery, mentioned later, and claims that nearly any RC transmitter and receiver combo will work with the hardware.

Finally, Parallax describes the quadcopter as requiring advanced-level skills to fly it, and it's suggested for experienced RC airplane operators.

FIGURE 3.1   The ELEV-8 stands out with its clean looks, with far fewer visible wires than is average (credit: Parallax).

FIGURE 3.2 The ELEV-8 is a difficult bird to fly (credit: Parallax).

The ELEV-8 has the following features. Feel free to check out the parts that make up the kit in Figure 3.3.

- **Frame**: Aluminum tubes with plastic plates and connectors.
- **Landing struts**: Plastic.
- **Motors**: Four 1000kV.
- **ESCs**: GemFan 30A.
- **Flight control**: The kit comes with a Hoverfly Open flight control board.
- **Power**: No batteries come with the kit, although a 3300 mAh, three-cell lithium polymer (LiPo) battery is recommended.
- **Cost**: $400–$550
- **URL**: http://www.parallax.com/product/elev-8

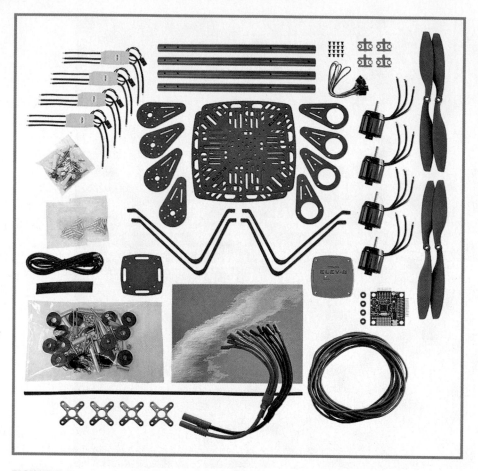

FIGURE 3.3 The ELEV-8 is a challenging but rewarding kit to build (credit: Parallax).

DJI Phantom 2 Vision+

The DJI Phantom is a pre-built, high-end quadcopter with a variety of configurations, ranging from $580 to $1,300. The top of the line is the Vision+ (pictured in Figure 3.4), which features a motorized gimbal (shown in Figure 3.5) so you can aim a camera wherever you want.

FIGURE 3.4 The DJI Phantom is good looking and ready to fly (credit: DJI).

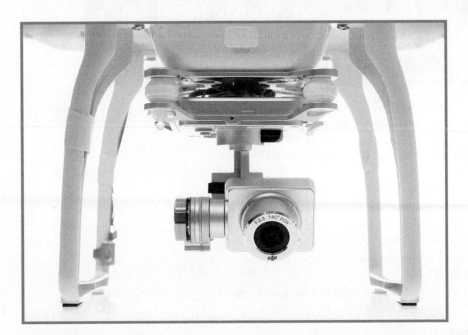

FIGURE 3.5 The Vision+, true to its name, has a gimbal-mounted camera that comes with it (credit: DJI).

The Phantom has far more control options than most quadcopters. The Vision+ comes with a 5.8GHz controller. This unit also has a smartphone holder so you can use a phone application to control the Phantom, or you can use the joysticks on the hand unit.

Finally, the Vision+ is very attractive, with a beautiful shell and LED strips on the booms, as shown in Figure 3.6.

FIGURE 3.6   The Phantom's distinctive LEDs not only illuminate the quadcopter, but also tell the operator which way is forward (credit: DJI).

DJI's Phantom 2 Vision+ has the following specs:

- **Frame**: Metal and plastic
- **Motors**: Four T-Motor MN2214 920kV brushless outrunners
- **ESCs**: Custom DGI controllers
- **Flight control**: DJI NAZA Autopilot
- **Power**: 5200 mAh LiPo battery
- **Cost**: $579–$1,229
- **URL**: http://www.firstpersonview.co.uk/quadcopters/dji-phantom-2-vision-plus

OpenROV

I've mentioned a category of drone called the remotely operating vehicle, or ROV. The best example of a DIY unmanned submersible is OpenROV, an open-source semipro ROV that has been tested in lakes and oceans around the world (see Figure 3.7). It has been used to explore caves, shipwrecks, and Antarctic glaciers. Less romantically, it has been used to inspect underwater features such as ship hulls that might otherwise require a professional diver.

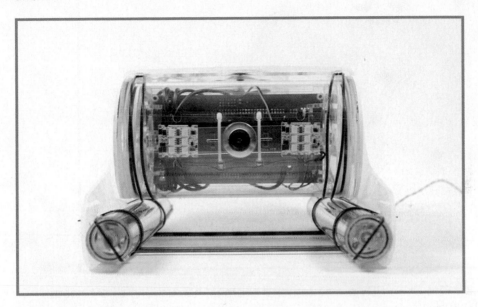

FIGURE 3.7 An OpenROV is a small submersible that explores underwater while the operator remains on the surface (credit: OpenROV).

OpenROV consists of three motors, with the control electronics sealed against the moisture. Because of the poor transmission of radio waves through water, operators must connect a laptop to the ROV with long wires (visible in Figure 3.8), which trail behind the craft as it explores underwater.

The operator looks through the attached webcam and moves the craft by interacting with a browser application. Additionally, because underwater tends to be dark, the ROV has a pair of LED matrices that it uses as headlights.

The OpenROV kit (shown in Figure 3.9) comes with the following parts:

- **Frame**: Laser-cut acrylic.
- **Motors**: Three brushless DC motors with Graupner high-efficiency marine propellers.
- **ESCs**: Three FalconSEKIDO brushless motor controllers.
- **Control**: OpenROV uses a BeagleBone Black that plugs into a custom Arduino Mega; an Ethernet add-on board handles communication with the surface.
- **Power**: Batteries not included, but it takes six "C" lithium cells.
- **Cost**: $899 kit, $1,450 assembled.
- **URL**: openrov.com.

FIGURE 3.8 The Open ROV's wire tether transmits instructions to the robot (credit: OpenROV).

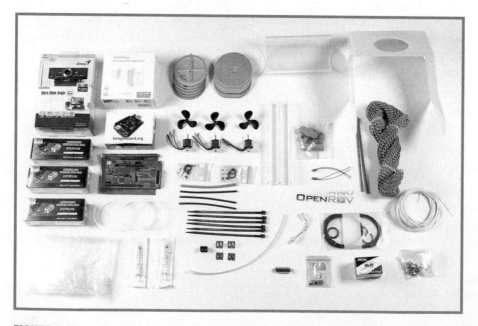

FIGURE 3.9 Put these parts together to make a submersible (credit: OpenROV).

Actobotics Nomad

Actobotics is a cool building set whose trademark element is the hole-studded beams you can see in Figure 3.10. Their newest robot offering is the Nomad, a rover just dying to be decked out in sensors and control systems. I mean this in the sense that you don't get a full kit with a microcontroller or control system—it's strictly DIY after you have the chassis built and add the motors and wheels.

FIGURE 3.10   The Nomad is a chassis and motors—just add microcontroller and battery pack!

The wheels are robust for a small robot. The largest chassis elements are a foot long. Each wheel is 5" in diameter and over 2" thick—just the ticket for clearing those obstacles. Check them out in Figure 3.11.

FIGURE 3.11 The Nomad features an intriguing "off-road" chassis.

That said, you will have to source your own battery packs, motor controllers, and so on. Do not buy the Nomad expecting it to roll right out of the kit, unless you have those extra parts! It's a great start for a rover, however, and the construction promises to be durable enough (see Figure 3.12) to explore any terrain your backyard has to offer.

The Nomad offers the following specs:

- **Frame**: Extruded aluminum frame with an ABS enclosure
- **Motor**: Four 12V, 313-RPM DC motors
- **Wheels**: All-terrain tires
- **ESCs**: None included
- **Control systems**: None

- **Power**: None
- **Cost**: $280
- **URL**: https://www.sparkfun.com/actobotics or https://www.servocity.com/html/actoboticstm.html

FIGURE 3.12   The knobby tires are a boon when traversing rough terrain.

Brooklyn Aerodrome Flack

Brooklyn Aerodrome (brooklynaero.com) is a group in New York making RC airplanes out of household insulation, the kind that looks like brightly colored foam—you can see an example in Figure 3.13. Equipped with an electric motor for propulsion and servos for manipulating the control surfaces, the Flack (which stands for Flying Hack) is extremely maneuverable. Because the airplane is made out of foam, rebuilding it is a cinch—if the plane crashes and the foam breaks, simply remove the electronics deck and put it on a new wing.

The Flack started off as a $100 kit, but now it's about twice that, with somewhat better electronics as well as a great book in which the group details what they've learned. If you're just getting into making inexpensive RC projects, this book will definitely help you along.

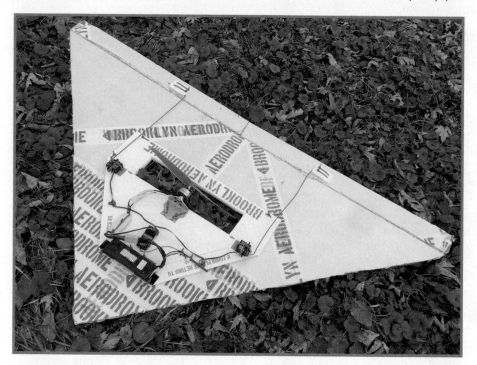

FIGURE 3.13   The Brooklyn Aerodrome Flack is an inexpensive RC plane anyone can build.

The kit (a partially-built Flack may be seen in Figure 3.14) provides everything you need to fly, including batteries, a base-model controller and receiver, motors, and enough insulation to rebuild the plane if it is wrecked. It makes adept use of lightweight materials like zip-ties (shown in Figure 3.15) but is sturdy enough to take a hit.

FIGURE 3.14   Available as a kit, the Flack must be assembled by the buyer.

The Flack features the following specs:

- **Frame**: Coroplast flight deck and foam insulation wings. You get a bunch of extra wings in the kit, or you can source your own.
- **Motors**: One 1800kV HiModel brushless outrunner; two TG9E micro-servos.
- **ESC**: No-name 18A ESC.
- **Props**: 9×9 slow flyer propeller.
- **Flight control**: Hobby King HK-T6A controller and receiver.
- **Power**: Turnigy 1800 mAh LiPo battery.
- **Cost**: $199 from brooklynaero.com or $249 from makershed.com.
- **URL**: http://www.brooklynaerodrome.com/.

FIGURE 3.15 The Flack makes adept use of lightweight materials like double-sided tape and zip ties.

Summary

In this chapter, we got in depth with five excellent drone products. Some are kits and others ship pre-assembled, but all offer their own unique challenges and learning opportunities. Now, having looked through a wide range of projects and kits, you're ready to tackle our own creation! In the next chapter, you'll begin working on your quadcopter, beginning with selecting an airframe.

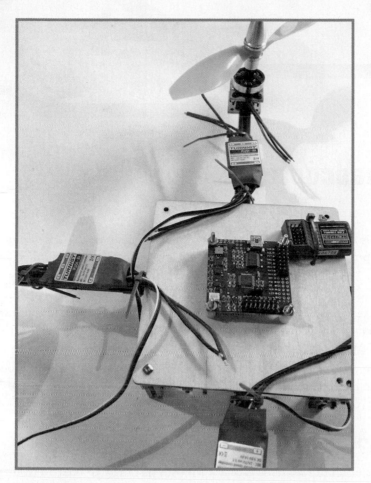

FIGURE 4.1 Build this quadcopter yourself!

For instance, the airframe pictured in Figure 4.2, from Parallax's ELEV-8 quadcopter, features lightweight aluminum tubes for the motor booms, with plastic mounting lugs designed specifically to mate with the motors and other components that come with a kit. It's to be expected, but it's kind of nice knowing everything will fit together.

This effortless compatibility and more polished appearance are a couple of advantages offered by commercial airframes. In the next section, we'll go over a number of features of these products to take into consideration when making a purchase.

Building a Quadcopter I: Choosing an Airframe

The main project of the book is a four-bladed helicopter (pictured in Figure 4.1) called a *quadrotor* or *quadcopter*. You'll begin the project by choosing a chassis, which in the plane world is called an *airframe*.

This chapter begins by presenting you with a number of chassis options, but ultimately (spoiler alert!) I went with a set of MakerBeam aluminum girders that I bolted together into a fine airframe. I then topped this off with a handy wooden platform that will eventually house the quadcopter's microcontroller, battery pack, and other electronics.

I'm getting ahead of myself, though! Before we get to the MakerBeam build, let's check out a bunch of other options, including commercial products and DIY possibilities. Once we check those out, I'll guide you through assembling your own MakerBeam airframe.

Which Airframe?

The funny thing about a drone or robot's airframe is that it's mostly just there to hold everything together, so consequently any reasonably rigid, strong, and lightweight material could be (and has been!) used to build a chassis. Sometimes this is done to hilarious effect, with all manner of odd things—recall some of the projects from Chapter 2, "Showcase of Cool DIY Drones."

There are wood airframes, plastic ones, and metal ones. If it's reasonably strong, light, and you can bolt stuff to it, chances are it will work as a chassis. That said, some airframes do offer considerable advantages.

FIGURE 4.2   The Parallax ELEV-8 features a lightweight airframe made out of plastic and aluminum.

Choosing Between Commercial Options

Let's go over the criteria one might consider in choosing an airframe. The following list discusses some of the features to take into consideration:

- **Appearance**—Anyone can *make* homely. If you're paying money for a chassis, it should look like it was designed and machined by professionals. It should look better than what you'd whip up in your basement.
- **Configuration**—How many motors will your copter feature? The number of motor booms is not the only configuration-related question to think about. Will you want to mount a camera on it? Depending on where you put the camera, you might need landing struts. The most common airframe is the now-classic quadcopter, featuring four motor booms with a central plate that supports the controller and batteries.
- **Dimensions**—How big of a quadcopter do you want? My Parallax ELEV-8 is over 2 feet across, and it's considered only typical by quadcopter standards. Keep the overall needs of your project in mind, as well as the technical specifications of your motors and props. Don't be hesitant to try out a smaller project first—the motors and other components may be cheaper because their technical requirements are less demanding.

■ **Material**—As I mentioned, pretty much any reasonably lightweight and sturdy material can be used for an airframe. That said, aluminum and plastic—or a combination of the two—are the most popular.

■ **Mounting hardware**—This one is huge for me. What use is a cool airframe if you can't easily bolt your components onto it? Wanting to have motors that easily bolt onto an airframe often means having specialized plates and attachments, although this isn't required. Many quadcopters have been built that are held together mostly with duct tape and zip ties.

■ **Price**—I see a big difference in prices, but sometimes it's not so apparent what you're getting for the extra dough. With all hobbyist hardware, there are some categories of product that have cool screen-printing on the housing and cost twice as much, but ultimately aren't all that impressive.

■ **Strength**—The dirty secret of quadcopters is that they crash—a lot! They're constantly plowing into the turf after batteries run out or a technical glitch occurs. How durable of a drone are looking to build? On the other hand, with strength often comes weight, and what good will it be to have an indestructible quadcopter that can't make it off the ground? Which brings us to...

■ **Weight**—The final criteria to consider is weight. The lifting power of your motors offsets the weight of the chassis, and if you have monster motors and props, you can get away with a more robust airframe.

Making Your Own Airframe

Although buying is always an option, it's definitely best to build an airframe if you have the time, the tools, and the materials. That way, you can have the perfect airframe for your needs, and you can take pride in having created something!

The following sections discuss the three basic ways to create your own airframe.

Building Set

With a building set, instead of designing anything, you simply build your airframe out of plastic or metal beams. Most DIY kits involve the bolting together of parts anyway—granted, custom parts rather than stock parts—but you can still see how easy it would be to build your own airframe.

In this chapter, I show you how to use a convenient and clever aluminum building set called MakerBeam to build a chassis, as seen in Figure 4.3.

FIGURE 4.3 Build an airframe just like this with the steps shown later in this chapter.

3D Printer

Another option is to print your own airframe using a 3D printer, a tool that creates three-dimensional objects out of melted plastic. There are already a bunch of quadcopter parts on Thingiverse, a site featuring 3D-printer files that can be freely downloaded. Take, for example, the T-6 Quadcopter, pictured in Figure 4.4. Its creator, Brendan22, designed and printed the booms and enclosure, and you can download his designs on Thingiverse at http://www.thingiverse.com/Brendan22/designs.

If you aren't content to download someone else's work, you can use 3D-design software such as SketchUp (sketchup.com) or Tinkercad (tinkercad.com) to build the part you need for your project, and then print it out on your handy 3D printer. If this sounds a little expensive, that's kind of true. 3D printing is a new industry, and prices haven't come down to the point where everyone has a 3D printer at home. Don't worry: There are plenty of other ways to build an airframe!

FIGURE 4.4 The T-6 Quadcopter has a 3D-printed body and six motors (credit: Brendan22).

Wood

Wood makes for a very lightweight and sturdy airframe material, especially for smaller and lighter quadcopters. A lot of model gliders use balsa, a super light and easily-shaped wood. However, quadcopters have the capability to carry a decent amount of weight, and that makes wood's relatively unimpressive strength-to-weight ratio less of a problem.

One fun aspect of wood airframes is that you can laser-cut the frame out of thin slats of wood and then piece them together like a puzzle. Figure 4.5 shows one example of this type of creation. Called the Flone (http://www.thingiverse.com/thing:113497), it's an airframe for a smartphone-controlled quadcopter. It looks great and is easy to make—if you have a laser cutter, that is.

Another advantage to wood is that it is a cinch to modify it on the fly—just drill a hole in it! Unlike commercial frames, or even metal and plastic ones, it's super easy to cut or drill into a wooden chassis. If you mess up, all you have to do is laser out another one!

FIGURE 4.5 The Flone airframe is easily laser-cut out of a piece of wood (credit: Lot Amoros).

Project #1: MakerBeam Airframe

For my quadcopter, I decided to make my own airframe, using some cool aluminum beams I had lying around. The beams, shown in Figure 4.6, bolt together very securely and connect to multiple angle plates so that the thing won't fly apart in midair.

MakerBeam

Called MakerBeam (www.makerbeam.eu), the beams are pretty cool, bolting together with M2.5 screws, which employ an unusual connection method—the heads of the screws are square, and they slide into grooves cut into the aluminum beams. Connector plates are added to the screws; then a hex wrench is used to tighten the nuts (see Figure 4.7).

FIGURE 4.6   The MakerBeam chassis serves as a light and flexible platform upon which to build your quadcopter.

FIGURE 4.7   MakerBeam's threaded end-hole and clever grooves make it useful for making a quadcopter chassis.

The product has a cool idea behind it. In 2012, a crowd-funding campaign launched OpenBeam with $100,000 in development money. The idea was to create an aluminum building set that was open source, so anyone could create their accessories or expansions on the base design.

MakerBeam is an offshoot of that original project, with different connectors and slightly modified beams, but still retaining the spirit of the original. In the U.S., you can buy MakerBeam on Amazon.com (search for the ASIN of B00G3J6GDM).

You can also buy the original OpenBeam (www.openbeamusa.com) from Adafruit. It works much the same way, but uses nuts trapped in the grooves, rather than the heads of screws. It also offers downloadable designs so you can output your own 3D-printable connector parts.

Parts

You'll need the following parts to build your airframe (shown in Figure 4.8). Note that all MakerBeam parts are found in the MakerBeam Starter Kit (P/N 01MBTBKITREG):

A. **Four 150mm beams** (P/N 100089).

B. **Four 100mm beams** (P/N 100078).

C. **Four 60mm beams** (P/N 100067).

D. **Eight corner brackets** (P/N 100315).

E. **Four right-angle brackets** (P/N 100326).

F. **Four L-brackets** (P/N 100304).

G. **M3 x 6mm screws** (P/N 100359), though they offer longer screws not found in the Starter Kit.

H. **M3 nuts** (P/N 100416). They also offer self-locking nuts (P/N 100405).

I. **A piece of wood**. I used a 13×13cm square of 3mm-thick (1/8th-inch-thick) Baltic Birch for the platform, with screw holes 11cm apart.

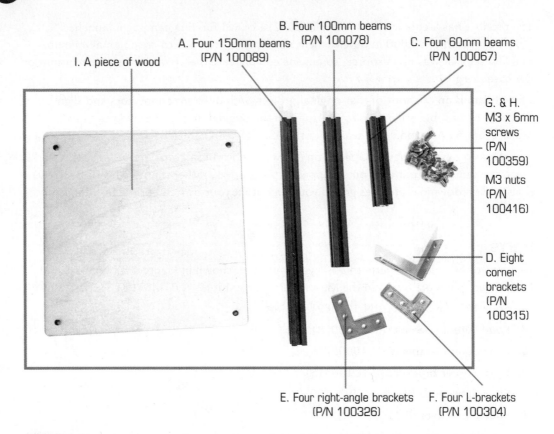

I. A piece of wood

A. Four 150mm beams
(P/N 100089)

B. Four 100mm beams
(P/N 100078)

C. Four 60mm beams
(P/N 100067)

G. & H.
M3 x 6mm
screws
(P/N
100359)

M3 nuts
(P/N
100416)

D. Eight
corner
brackets
(P/N
100315)

E. Four right-angle brackets
(P/N 100326)

F. Four L-brackets
(P/N 100304)

FIGURE 4.8 You'll need these parts to build your MakerBeam airframe.

Steps

Once you have gathered all your parts together, follow along with these steps to build your airframe:

1. Make four identical assemblies, each consisting of a motor strut with a section of the central square. These sub-steps show how to make each one:

   a. Slide two screws into the groove of a 150mm beam. Secure a right-angle bracket to those two screws using the supplied nuts and hex driver, as shown in Figure 4.9. (Note that I only tightened every other screw, so I could make adjustments more quickly. I'll go through later on and finish adding nuts once the design is the way I like.)

   b. Slide two screws into the groove of a 10cm beam. Connect it to the 15cm beam and bracket you already prepared, as shown in Figure 4.10. Secure it with two nuts.

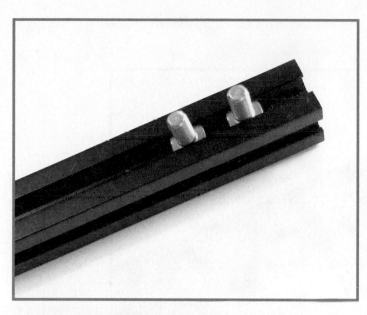

FIGURE 4.9 Slide the heads of two M3 screws into the grooves of a 150mm beam.

FIGURE 4.10 Make a "T" with the two beams and secure with a bracket.

c. Add a corner-bracket to help secure the beams, as shown in Figure 4.11. Secure the bracket the normal way.

FIGURE 4.11 Reinforce the "T" with another bracket.

d. Attach an L-bracket to the top of the 10cm beam, positioning it like you see in Figure 4.12.

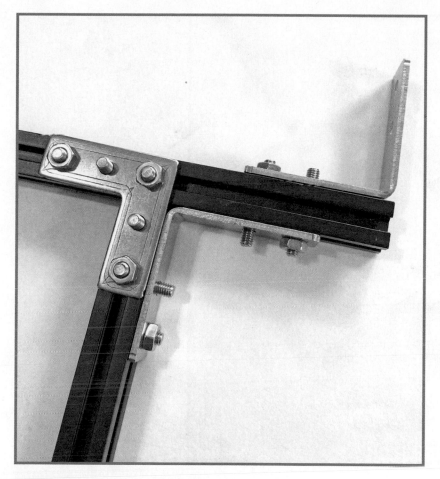

FIGURE 4.12 Add another bracket.

2. Once you have all four built, connect them together using the usual hardware. It should look just like Figure 4.13.

FIGURE 4.13   Add the four segments together, and you start seeing your airframe take shape.

3. Flip over the airframe so the flat L-brackets are on the underneath. Add four 6mm beams and secure them with L-brackets. Figure 4.14 shows how it should look.

4. Now you're ready to add the wooden platform, shown in Figure 4.15. I laser-cut the precise shape I needed, but you can use any old (thin) piece of wood and hand-drill the holes. Don't make it too thick! Baltic Birch no thicker than 1/8" (3mm) does the trick. Screw the platform into the threaded holes in the tops of the four legs using M3 bolts, secured with a hex wrench.

FIGURE 4.14 Add legs!

FIGURE 4.15 Attach the wooden platform to the tops of the legs.

Does your quadcopter seem incomplete? It should, considering that there are no motors, props, or electronics. Be patient! In future chapters, you'll have an opportunity to complete the build.

Summary

You're well on your way to building your very own quadcopter—you constructed the airframe out of aluminum beams! In the following chapters, you'll add the motors and props, battery pack, and microcontroller. But let's mix it up! In Chapter 5, "Rocket Drone Project," you'll build an electronic payload that will datalog the G-forces experienced by a model rocket in fight.

5

Rocket Drone Project

Let's mix things up by exploring a different type of drone. I'm talking rocketry, which could be described as a type of fully autonomous drone, if you think about it. In this chapter, you learn a little about the history of hobby rocketry and go through the steps to actually build a data-gathering rocket drone that records its altitude so you can check it out later. You can see the rocket in Figure 5.1.

FIGURE 5.1 You'll build the data-gathering rocket this chapter.

Amateur Model Rocketry

People have been playing around with rockets since they were invented centuries ago. In the United States today, this mostly involves small plastic-and-balsa rockets manufactured by Estes Rockets, a Penrose, Colorado company. Estes sells solid rocket motors and rocket kits (such as the one shown in Figure 5.2) as well as launch systems.

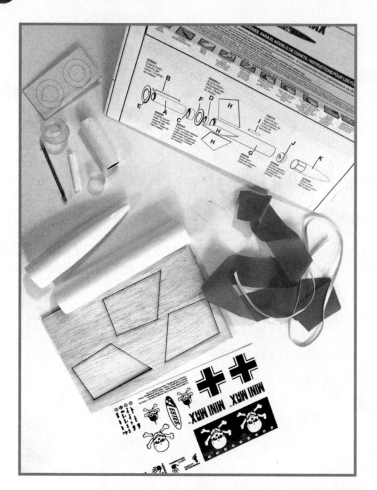

FIGURE 5.2   An Estes rocket kit. Just add glue and paint!

In 1959, Vern Estes developed a machine that packed solid rocket engines, and they created rocket kits around it. Inexpensive and just challenging enough, the Estes rockets soon became a worldwide hit. Estes rocket engines are essentially solid propellant in a cardboard tube, and they contain a secondary charge to deploy the parachute once the main charge burns out. The company also sells launch systems that consist of a handheld controller and a launch pad with a slender metal rod that guides the rocket skyward.

Estes rockets range in size from very small rockets to huge ones. At the low end is the minuscule Mosquito, which is so light that it tumbles down safely without a parachute. At the opposite end of the spectrum is the Leviathan, packing a powerful engine that propels the rocket to altitudes of over 1,500 feet. There are a lot of different designs, and you can even buy a variety pack of rocket tubes and nose cones to create your own.

Let's go over the anatomy of a typical model rocket, following along with Figure 5.3:

A. Nose cone—Often the only plastic part on a model, the nose cone helps make the rocket more aerodynamic, while being tough enough to take the impact if it hits the ground hard on the way down.

B. Shock cord—This is merely the cord (in this case, a rubber band) that connects the ribbon/parachute, the nose cone, and the rocket body together.

C. Ribbon or parachute—After the rocket engine burns out, it triggers a secondary charge that deploys the parachute—or in this case, a ribbon. Smaller rockets don't need a parachute and get by fine with a ribbon, which provides just enough drag to slow down the rocket's velocity so it doesn't break when it hits the ground.

D. Recovery wadding—This fire-resistant wadding protects the parachute from the secondary charge (not shown in Figure 5.3).

E. Body—The (usually) cardboard tube forms the central body of the rocket.

F. Fins—These provide stabilization. In the Estes world, they're usually laser-cut balsa. Most kits assume you'll be sanding and painting the fins and then gluing them onto the tube. This is actually rather difficult to do right the first time.

G. Launch lug—This tube guides the rocket up the metal rod that is part of the launchpad. It looks an awful lot like a length of drinking straw.

H. Rocket motor—Estes motors consist of a cardboard tube with propellant and a secondary charge packed inside.

I. Igniter—This wire heats up when electricity passes through it, thus igniting the rocket fuel.

J. Plug—This plug keeps the igniter from falling out prior to launch.

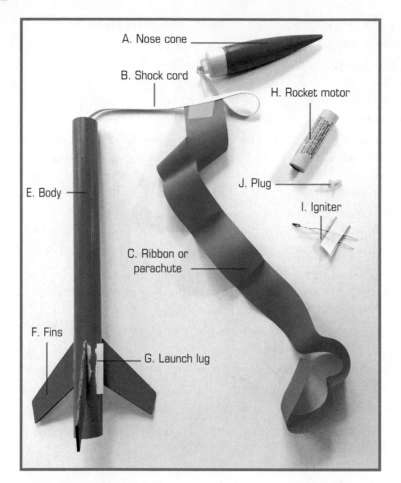

A. Nose cone

B. Shock cord

H. Rocket motor

E. Body

J. Plug

I. Igniter

C. Ribbon or parachute

F. Fins

G. Launch lug

FIGURE 5.3 Though often rather small, a model rocket packs a lot of parts.

Quick-and-Dirty Arduino Guide

Now that you're up to speed on rocketry, let's bone up on another skill you'll need to make this chapter's project: the Arduino microcontroller.

This chapter's project involves an Arduino, an easy-to-use microcontroller that manages our data-gathering rocket payload. The following is a simple how-to describing the process of uploading a program (called a *sketch* in the Arduino world) to the board.

The project in this chapter uses an Arduino Micro (see Figure 5.4), an Arduino small enough to fit inside the rocket's tube. Nevertheless, the Micro can do pretty much the same things as the full-size model, making it ideal for our purposes.

FIGURE 5.4   The Arduino Micro is a compact Arduino useful for small-space tasks.

You'll also need a USB cable. The exact type of cable depends on your Arduino. A Micro uses a micro-USB cable (Sparkfun P/N 10215) whereas an Arduino UNO uses a standard USB A-B (P/N 512). You can find out more about cable selection at Arduino.cc. You'll also need a reasonably up-to-date desktop or laptop computer—it can be a PC, Mac, or Linux.

Once you have the equipment you need, follow these steps:

1. Download and install the Arduino software. You can download the software and read detailed instructions at Arduino.cc. A screenshot of the website is provided in Figure 5.5.

2. Launch the Arduino software and plug in your Arduino via a USB cable, as shown in Figure 5.6.

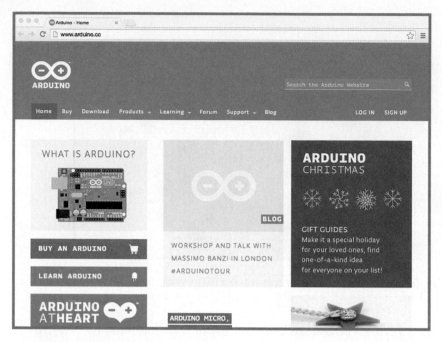

FIGURE 5.5 The mother lode of Arduino lore, Arduino.cc should be your first destination for learning about Arduinos.

FIGURE 5.6 Plug in your Arduino.

3. Go into Tools, Boards and select your Arduino from the list, as shown in Figure 5.7.

FIGURE 5.7 Select your Arduino from the list.

4. You'll also have to select a port. That is in the Tools menu as well, as shown in Figure 5.8. You may have to try a couple different ports to find one that works.

FIGURE 5.8 Choose the port.

5. Now select File, Open to open up your "sketch," which is what programs are called in the Arduino world. You can either download someone else's sketch from the Internet, or use an example sketch such as Blink, which is found under File, Examples, Basics, Blink (see Figure 5.9). Blink is sort of the "Hello World" of hardware hacking. It's the first thing you learn!

FIGURE 5.9 Open the Blink sketch.

6. Click the Upload button and send the sketch to the board, as shown in Figure 5.10. You're done! The Arduino is now programmed and will launch the sketch automatically when it is powered on. You're ready to build your rocket!

If an error message crops up, double-check your settings and try a different port. If all else fails, try the support forums and FAQs on Arduino.cc.

FIGURE 5.10 Upload the sketch to the Arduino.

TIP

Arduino for Beginners

If you'd like to learn more about Arduinos, be sure to check out *Arduino for Beginners* (Que, 2013), a book of Arduino projects, including sketches and tool tips (see Figure 5.11). It teaches a bunch of hardware and software concepts that will help you level up your Arduino knowledge.

FIGURE 5.11 If you're an Arduino beginner, get this book!

Project #2: Data-Gathering Rocket

Now that you're up to speed on Arduinos, let's tackle building a data-gathering rocket (shown in Figure 5.12) with an altimeter and the ability to log its data so you can study it afterwards. We'll use an Estes V2 rocket, which is big enough to fit a circuit board and battery inside. I'll show you how to wire it up and program the Arduino rig that will be the brains of the rocket.

FIGURE 5.12   The data-gathering rocket records its telemetry for later study.

Parts for Building the Data-Gathering Rocket

You'll need the following parts to build your data-gathering rocket:

- **Estes semi-scale V2 model rocket**—You can get this from most hobby stores or Estesrockets.com (P/N 003228).
- **Estes Porta-Pad II launch system**—This is just a glorified 9V battery to set off the engine (Estesrockets.com; P/N 002215).
- **Arduino Micro**—Buy it at Adafruit.com (P/N 1086) or another online store.

- **Altimeter**—There are tons of altimeters and accelerometer breakout boards around, but I chose the MPL3115A2 (Adafruit.com; P/N 1893). It determines altitude changes by sensing barometric pressure.
- **OpenLog datalogger**—Sparkfun sells these (P/N 9530).
- **Half-size breadboard**—This plastic breadboard (Sparkfun.com; P/N 12002) is probably not the lightest way to go.
- **Jumpers**—Also known as "wires." Adafruit.com has a good set (P/N 153).

Steps for Building the Data-Gathering Rocket

Once you have your parts together, you can begin building the project. Just follow these steps:

1. Build the rocket. Follow the instructions that came with the V2, and assemble and paint it as usual. You can see my rocket in progress in Figure 5.13.

2. Plug your Arduino into the breadboard as shown in Figure 5.14.

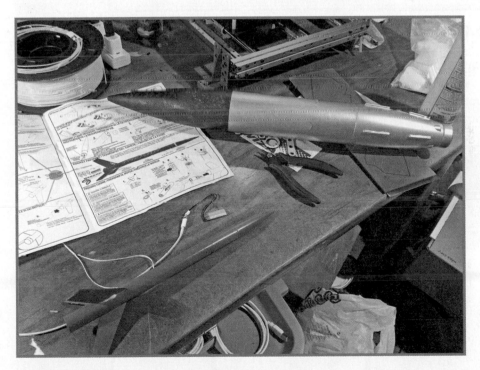

FIGURE 5.13 Assemble the rocket and then paint it!

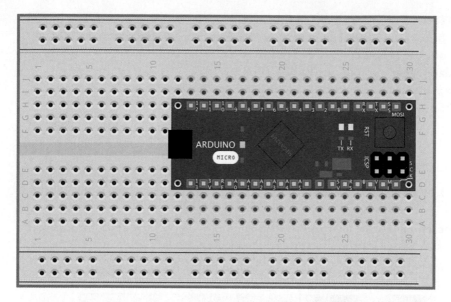

FIGURE 5.14 Insert the pins of your Micro into the breadboard.

3. Plug in the datalogger module and wire it up as shown in Figure 5.15. VCC plugs into 5V on the Arduino, while BLK plugs into the Micro's ground. (Note that the datalogger has both a GND and BLK pin, but leave GND alone!) Finally, connect the RXI pin on the OpenLog with the TX pin on the Arduino.

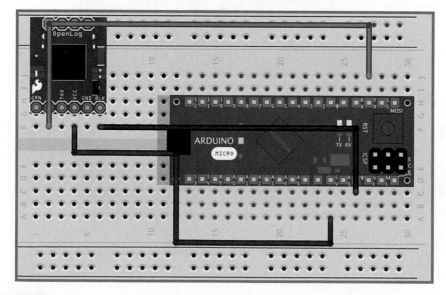

FIGURE 5.15 Install and wire up the datalogger.

4. Plug in the altimeter as shown in Figure 5.16. SDA on the OpenLog is connected to pin 2 of the Micro, shown as a yellow wire. SCL (the green wire) on the OpenLog connects to pin 3. Power and ground connect as shown.

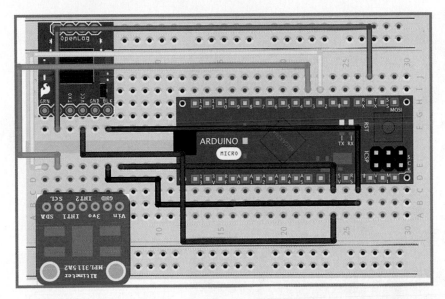

FIGURE 5.16 Next, add the altimeter and its wires.

5. Add the battery. The leads plug into the Micro pin marked VI (voltage in) as well as ground, shown in Figure 5.17.

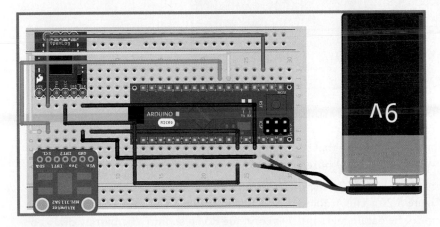

FIGURE 5.17 Connect the battery's leads to power your project.

6. Load the Arduino sketch from Examples and upload it to the Arduino, as shown in Figure 5.18. I'll describe this step more in the following section, "Programming the Payload."

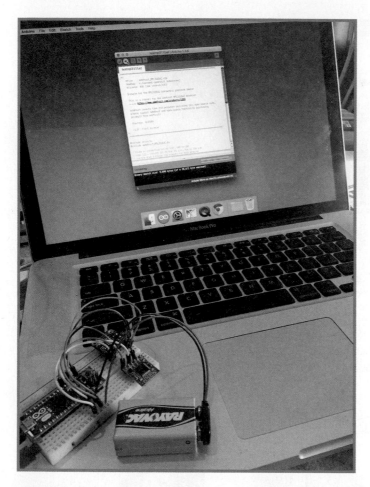

FIGURE 5.18 Upload the sketch to the Arduino.

7. Add the payload in such a way that it doesn't overbalance the rocket (see Figure 5.19). My solution was to cut into the nose cone and hot-glue the breadboard in place.

8. When you're ready to launch, power on the Arduino by switching on or plugging in the battery (see Figure 5.20). The Arduino is now taking readings from the altimeter and recording them to the datalogger, and will continue doing so until the battery drains or is disconnected.

FIGURE 5.19 Add the payload to the rocket.

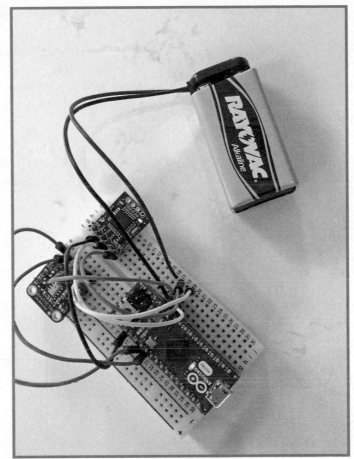

FIGURE 5.20 Power on the Arduino to start logging data.

9. Once the cataloguer has been recovered, power off the Arduino and insert the card into a reader to see the files.

Programming the Payload

You'll have to take my word for the fact that this code is about as simple as it gets. The code you need is simply the example sketch from Adafruit's MPL3115A2 accelerometer. The datalogger works seamlessly with it, so you don't have to modify anything. Here's what you do:

1. Download the MPL3115A2 library from https://github.com/adafruit/Adafruit_MPL3115A2_Library. Libraries are add-on code that is stored in a separate file from

your main sketch, allowing you to keep your code as clean and succinct as possible. Click the Download ZIP button to the right of the screen.

2. Follow the directions in http://arduino.cc/en/Reference/Libraries to install the library. The process is quite simple: Open up the Arduino folder on your computer and find the subfolder labeled Libraries. Uncompress the Adafruit library (this might require renaming the folder, depending on your operating system) and put the folder inside the Arduino, Libraries folder.

3. Restart your Arduino software and find the testmpl3115a2 sketch under File, Examples. However, let's go over the code to ensure that we know what's going on:

```
//these are the libraries you need in order to run this sketch. The Wire.h one
comes with Arduino so don't worry about it.
#include <Wire.h>
#include <Adafruit_MPL3115A2.h>

Adafruit_MPL3115A2 baro = Adafruit_MPL3115A2();

void setup() {

//the next two lines establish the serial connection and begin with a test mes-
sage.
  Serial.begin(9600);
  Serial.println("Adafruit_MPL3115A2 test!");
}

void loop() {
//this loop runs indefinitely as long as the Arduino is getting power.

  if (! baro.begin()) {
    Serial.println("Couldnt find sensor");
    return;
  }

//the altimeter takes a barometric reading
  float pascals = baro.getPressure();
  / Serial.print(pascals/3377); Serial.println(" Inches (Hg)");

//the altimeter determines altitude
  float altm = baro.getAltitude();
  Serial.print(altm); Serial.println(" meters");
```

```
//the altimeter has a little temperature sensor  in it.  Why not?
   float tempC = baro.getTemperature();
   Serial.print(tempC); Serial.println("*C");

   delay(250);
}
```

Summary

In this very exciting chapter, you got to build a rocket drone that records its altitude and acceleration during launch. In Chapter 6, "Building a Quadcoptor II: Motors and Props," you'll continue the main quadcopter project by selecting and adding motors and propellers to the chassis you already built.

6

Building a Quadcopter II:
Motors and Props

Next up in our project to build a quadcopter (Figure 6.1), you'll tackle motors and propellers. I'll guide you through choosing both, then I'll add my choices to the drone we are building. In Chapter 4, "Building a Quadcopter I: Choosing an Airframe," I showed how to build the airframe out of MakerBeam. With the addition of motors, props, and the mounting hardware for both, we'll be well on our way to getting airborne.

FIGURE 6.1 In this chapter, you'll add motors and propellers to your quadcopter.

Choose Your Motors

When buying quadcopter motors, you have dozens (if not hundreds) of models to choose from. That said, there are a number of ways to categorize DC motors. Let's examine the options.

Outrunner Versus Inrunner

You'll often see the terms *outrunner* and *inrunner* in the RC (radio control) world. These refer to the physical design of the motor housing. An outrunner (such as the one shown in Figure 6.2) rotates the entire housing; there is no rotor in the usual sense. Instead, the propeller is bolted onto the housing. Outrunners are often used in quadcopters because they have the ability to turn big props very well. On the downside, outrunner almost never come with a gearbox, thus limiting your ability to dial in the speed and torque that's right for your project.

Inrunners are the classic motor we think of when we hear the word. A ring of electromagnets rotates a ferrous rod. Often an inrunner will feature a gearbox, so you'll have a lot of options with regard to RPM (revolutions per minute) and torque.

FIGURE 6.2 Outrunner motors have a fixed base around which the entire casing rotates.

Brushed Versus Brushless

These are terms you're likely to encounter while motor shopping. It refers to the way that the magnetic coils inside the motor are energized. The brushed motor uses little metal brushes that touch the coils, which are wrapped around the rotor—the part of the motor that turns. In the brushless motor, it is the coils that are static, so no brush is needed.

There are advantages to both types. Brushless motors have better heat dissipation and therefore can be built more compactly. On the other hand, you can't just plug them in; they need relatively complicated control systems. By contrast, regular-old brushed motors (such as the one pictured in Figure 6.3) can be run with just a battery and don't need a controller. The downside that comes with brushed motors is that those brushes wear out.

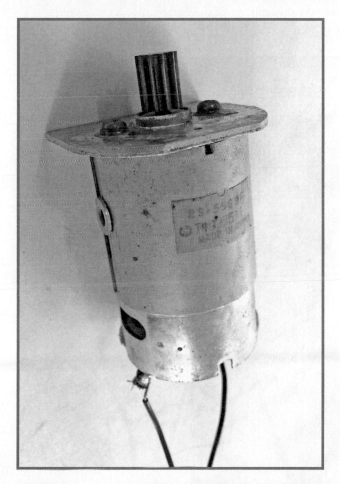

FIGURE 6.3 A brushed motor, identifiable by its two leads.

AC Versus DC

You know the difference between AC and DC, right? Alternating current is most commonly encountered in household outlets, though some hobbyist electronics uses it. Most relevantly, many drone motors are AC. By contrast, batteries and most electronic sensors and other modules operate on DC, or direct current.

The main reason you need to know this is that you must match your ESCs (electronic speed controllers) to the motor. If it's an AC motor such as the one in Figure 6.4, you'll need an AC ESC. Don't know anything about ESCs? Never fear, in Chapter 8, "Building a Quadcopter III: Flight Control," I'll tell you all about them.

So you're probably wondering how a DC battery can power an AC motor. The answer is that ESCs have inverters that translate the raw voltage of the battery into a three-phase signal that tells the motor how fast to turn.

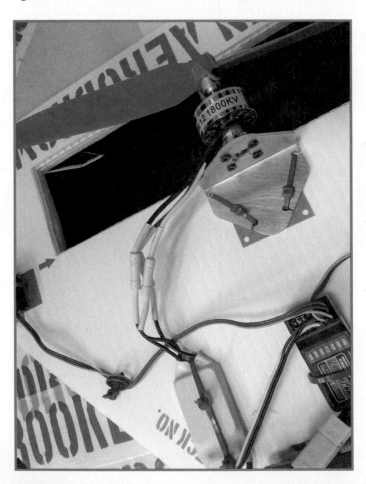

FIGURE 6.4   An AC motor with the matching speed controller.

Choose Your Propellers

Propellers are inexpensive and often broken, so you'll have many opportunities to try out different types. In the meantime, here are some basic tips for choosing props for your quadcopter:

■ Propellers are measured in terms of diameter and pitch. Pitch refers to the angle of the propeller blade. The propellers I use in this book are 7×3.8.

■ You will often encounter two kinds of propellers sold as a set. These are regular and "pusher" propellers, as shown in Figure 6.5. The pushers are made to be rotated counterclockwise, whereas regular props rotate clockwise. These counter-rotations help stabilize the quadcopter.

■ Make sure you're buying slow-fly props, used for slow-flying electric aircraft such as quadcopters. Contrast this with smaller props used for fast-moving planes.

■ You may want to make two adjacent propellers a different color. This will help you iden-tify the front of the quadcopter, which is helpful for steering! Confusingly, many pusher props are given a different color, which doesn't help because they're mounted diagonally to each other.

FIGURE 6.5 Regular and pusher propellers are made to rotate opposite each other.

Prop Adapters

Another aspect of propellers to consider when choosing a set is, how will you install them? You'll often find the right kind of adapter sold alongside the propellers, but be sure you get the right kind of prop for the hardware and the right hardware for the motor. The relevant stat is the rotor diameter, so double-check this before you buy.

However, before you grab the first thing you see, be aware that there are two main schools of thought when it comes to attaching a propeller to a motor: collets and prop savers.

- Collets are metal clamps that look like cones—you can see an example in Figure 6.7 in the following section. These adapters are very secure, and you can be sure your propeller won't fall off the motor shaft. Surprise, this is not necessarily a good thing! It turns out that quadcopters break propellers all the time. Think of it—the props are thin plastic rotating at high speeds. If the propeller hits a sidewalk or wall, it's going to break. Many a drone-flying outing has ended abruptly when the pilot ran out of propellers! I used this method for the quadcopter project in this book.
- Prop savers (shown in Figure 6.6) are another way of attaching the propeller—and true to the name, the idea is to avoid damaging so many propellers in crashes. The way it works is that the propeller is held on with just a rubber band, counting on friction and acceleration to keep it secure. However, if the quadcopter crashes, the propeller simply falls off and can be reattached. If you go this route you'll need lots of extra rubber bands, and be sure to check the props before every flight.

FIGURE 6.6 Prop adapters connect the propeller to the motor.

Project #3: Attach the Props and Motors

For the next segment of the quadcopter project, you'll tackle adding the motors and propellers, as shown in Figure 6.7. Each boom of the quadcopter will get a motor, a propeller, and the hardware to mount both. So without further ado, let's get started.

FIGURE 6.7 You'll attach the motors and propellers to your quadcopter.

Parts

You'll only need a few parts for this step of the project:

- **Four motors**—I used Hobby King 1400kV brushless motors (P/N 2205C-1400).
- **Four props**—I used 7×3.8 Turnigy slow-fly electric props (Hobbyking.com; P/N 9329000203-0). You'll want two pusher props as well (P/N 9329000206-0).
- **Four prop adapters**—Use a Hobby King "collet-type" prop adapter for 3mm rotors (Hobbyking.com; P/N GON-D3T6).
- **Mounting plates**—You can either laser out the mounting plates I designed (http://www.thingiverse.com/jwb/designs) or 3D-print some equivalent lugs (http://www.thingiverse.com/thing:198878). The laser-cut mounts are made from 1/8" plywood and need #4 screws (0.75") and nuts to tension them.

Steps for Attaching the Props and Motors

Follow these steps to attach the motors and propellers to your quadcopter airframe:

1. Print or laser cut the motor mounts. In the parts list, I mentioned two different ways to make mounts for the motors. Because my 3D printer wasn't working when I worked on this step, I decided to laser cut mounts. You can see what I came up with in Figure 6.8. If you don't have access to a laser cutter, I'd suggest making similar plates using 1/8" plywood, cut and drilled to match the laser pattern. Finally, you can buy motor mounts online—just make sure they're compatible with a 1cm beam.

FIGURE 6.8   Cut out the motor mounts using a laser cutter.

2. Attach the motor mounts to the booms, as shown in Figure 6.9. I bolted the wooden parts together using #4 screws for tension. If you need more friction, try trapping a piece of double-sided tape between the wood and metal.

FIGURE 6.9 Attach the motor mounts to the booms.

3. Attach the motors to the mounts using M2 screws and nuts, shown in Figure 6.10. If you can't get locknuts, be sure to use a thread-locking solution to help secure those motors.

FIGURE 6.10 Attach the motors to the mounts.

4. Connect the propellers to the motors using the prop adapters. You can see how to do this in Figure 6.11. All you do is thread the prop on the hardware while attaching the female end of the adapter to the motor's hub. Tighten until they feel secure, but not so tight that you'd have difficulty removing the prop if it breaks. Make sure you use the right propellers, with the two pushers opposite each other, and the two normal props opposite each other.

FIGURE 6.11 Attach the propellers.

Summary

You've made very important progress on the quadcopter project, learning about motors and propellers and adding them to the drone's chassis. In Chapter 7, "Blimp Drone Project," you'll put this knowledge to the test by building a drone out of toy balloons, a couple of motors, and a servo, and using two different ways of controlling it.

Blimp Drone Project

In this chapter, you'll check out radio control (RC) technology, which lets you remotely operate model vehicles using a wireless handheld controller. After that, you'll build a blimp drone that uses an RC setup to fly around (see Figure 7.1). Finally, I'll show you how to make the blimp run autonomously using an Arduino as the brains.

FIGURE 7.1   You'll build a blimp drone in this chapter.

Radio Control

Typical RC setups include three major control components: a transmitter, a receiver, and a speed controller for each motor. Let's check out each one.

Transmitters

RC setups consist of a handheld controller (shown in Figure 7.2) that includes a transmitter that sends the signals out to the model. These are control elements such as joysticks and switches. Using the controller, you can steer the model, adjust speed and flaps, turn on servos, and so on.

Not every transmitter is the same, of course. Free-spending hobbyists hanker after thousand-dollar pieces of equipment that are likely not worth anything close to that, but look impressive nonetheless.

More practically, higher-end transmitters do have better features, including more channels (the number of motors or other powered parts that can be controlled) and posh add-ons such as LCD screens and big antennas.

Most of the time, however, you don't need to spend a lot of money to get something that will steer your quadcopter. A basic transmitter and receiver combo can be purchased for well under $25.

FIGURE 7.2 A cheap transmitter such as this Hobby King model is a great entry into RC.

Receivers

The model needs a receiver to hear the radio signals and interpret them. A typical low-end receiver is shown in Figure 7.3. It consists of a series of contacts for controlling motors, an input to power the receiver's logic, along with an antenna of some sort to pick up the signals.

As mentioned, you'll often find receivers paired with transmitters, for obvious reason: No tinkering! You can be assured that the combo you buy will work "out of the box."

Receivers are differentiated by signal frequency (for example, 2.4GHz), number of channels, as well as style of antenna. Needless to say, these should match the transmitter you're using.

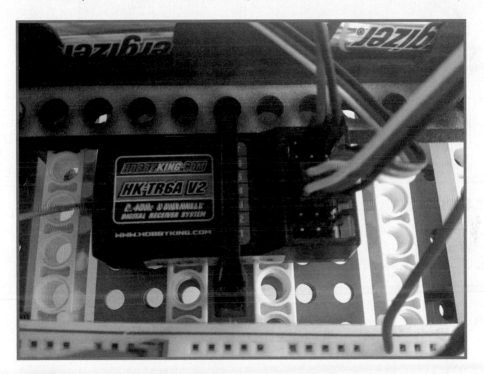

FIGURE 7.3 A Hobby King receiver interprets signals from the transmitter.

ESC (Electronic Speed Controller)

The voltage output by the receiver is too low to drive a motor, so it's used to trigger an ESC (shown in Figure 7.4) that manages the battery's full voltage. In addition, ESCs often have a microchip with certain actions programmed in. Some examples include braking, throttle range, and low-speed startups so the quadcopter doesn't zip off somewhere the instant the motors power on.

To choose an ESC, select one with an amps rating appropriate for the motors you're using—the general rule is to slightly exceed the highest rating of the motor. Similarly, be sure to get the right ESC for the type of motor, paying attention to AC versus DC and brushed versus brushless.

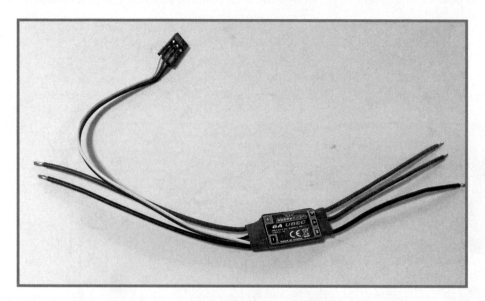

FIGURE 7.4 ESCs control motors with the help of the receiver.

Project #4: Blimp Drone

This chapter's project is a blimp (see Figure 7.5) that uses Mylar balloons to lift a wooden gondola into the air. The gondola consists of a pair of propellers that are angled by a servo, and you'll use either RC or an Arduino to control them.

FIGURE 7.5 The blimp drone uses balloons to hoist itself into the air.

Parts

You'll need the following parts to build the blimp drone. Note that the motors, propellers, and hardware are the same as the ones you used in Chapter 6, "Building a Quadcoptor II: Motors and Props."

- **Two motors**—I used Hobby King 1400kV brushless motors (P/N 2205C-1400).
- **Two prop adapters**—Use Hobby King "colet-type" prop adapter for 3mm rotors (P/N GON-D3T6).
- **Two propellers**—I used 7×3.8 Turnigy slow-fly electric props, one normal and one pusher (Hobbyking.com; P/N 9329000203-0 and 9329000206-0).
- **Two Mx10 hex screws to secure the motors (Hobbyking.com; P/N HA0506)**—You'll also need hex nuts (P/N OR017-01001-M2).
- **Servo**—I used a Hitec HS322HD servo (Jameco.com; P/N 395760).
- **Servo arm**—I used an Acrobatics single-side servo arm (P/N 525116).
- **Servo standoffs**—Use 1" female-to-female #6 standoffs (Allelectronics.com; P/N SP-263). You'll also need #6-32 screws to secure them.
- **Timing belt**—Adafruit.com sells one (P/N 1184).
- **Zip ties**—Just whatever you have lying around!

- **Laser-cut gondola**—You're welcome to download and use my design (http://www. thingiverse.com/jwb). For best effect, use acrylic, 1/8" birch, or some other similarly lightweight material.
- **Dowel (0.25")** —Use a length of around 18".
- **Helium balloons**—I used 24" Qualatex bubble balloons for their large gas volume.
- **RC transmitter and receiver combo**—Check Hobbyking.com (P/N HK-T6A-M2).
- **Two ESCs**—Use 6A Hobby King ESCs (Hobbyking.com; P/N 261000001)
- **Battery**—Use a Turnigy Nano-Tech LiPo battery, rated for 460 mAh (Hobbyking.com; P/N N460.3S.25).

Arduino Parts

If you're going to use an Arduino to control the blimp, this is what you'll need in place of the RC transmitter and receiver:

- **Arduino UNO or Micro**—I used a Micro in Chapter 5, "Rocket Drone Project," but UNOs are much more common. In the project description, I'll show you how to use either.
- **Two ultrasonic sensors**—I used PING-compatible sensors (Jameco.com; P/N 2206168).
- **Wire**—I used jumpers from Sparkfun.com (P/N 11026).

Steps

Follow these steps to build your blimp:

1. Assemble the gondola box shown in Figure 7.6. I cut the parts out of 1/8" birch using a laser cutter and then glued and clamped the parts. However, you don't need to get that fancy. Any lightweight box will do.

FIGURE 7.6 Assemble, glue, and clamp the chassis box.

2. Add the axle. Slide it through the gondola so about six inches of dowel protrude from either side. Use wood glue to secure the wooden washers (part of the laser-cutter design) to the dowel. You can see how the washers should look in Figure 7.7.

FIGURE 7.7   Use wooden washers to secure the dowel.

3. While the washers are drying, assemble and glue the motor mounts. You can see one of them in Figure 7.8—you'll need two, one for each motor.

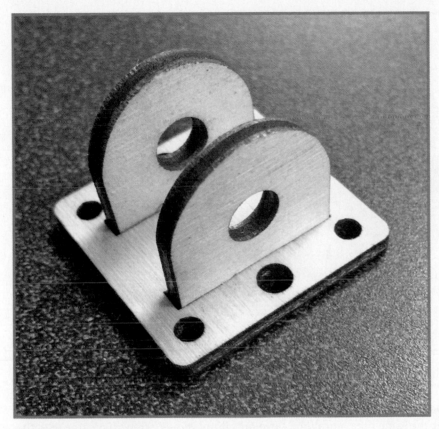

FIGURE 7.8 Assemble and glue the motor mounts.

4. Once all the glue is dry, slide the motor mounts onto the dowel and glue them into place, as shown in Figure 7.9. It probably goes without saying that the mounts should be oriented in the same direction.

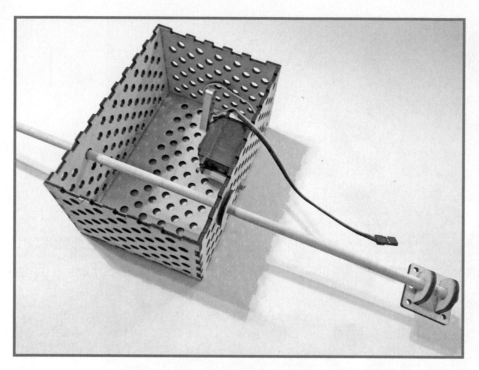

FIGURE 7.9   Glue the motor mounts into place.

5. While the motor mounts are drying, install the servo using the #6 standoffs and hardware. At the same time, install the servo arm using the set screw that came with the servo. Figure 7.10 shows how it should look.

6. Attach the motors to the mounting plates using the M2 screws and nuts. You may want to use some sort of thread-locking solution or locknuts to secure the screws. Once the motors are installed, gravity should rotate them so they're facing straight down, as shown in Figure 7.11.

7. Attach the propellers using the propeller adapters, just as you did in Chapter 6. When you're done with this step, the build should look like Figure 7.12.

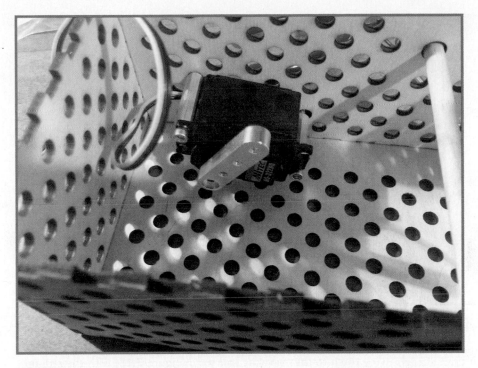

FIGURE 7.10 Install the servo and servo arm.

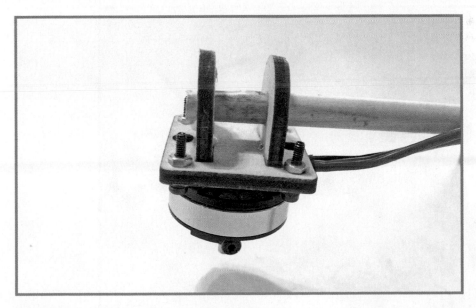

FIGURE 7.11 Next, install the motors.

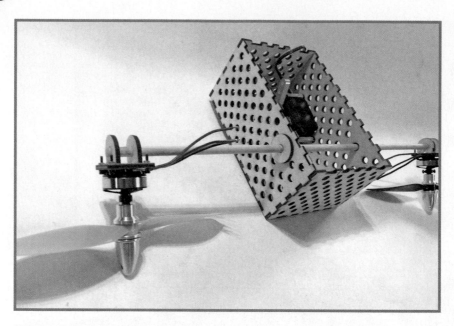

FIGURE 7.12 The propellers and prop adapters are next.

8. Attach the ESCs, zip-tying them to the dowel as you see in Figure 7.13, with the red and black power wires toward the gondola and the red-blue-black wires near the motors.

FIGURE 7.13 Zip-tie the ESCs to the dowel.

9. Use zip ties to attach your LiPo battery to the gondola, as you see in Figure 7.14.

FIGURE 7.14 Next, secure the battery with zip ties.

10. Connect the receiver to the gondola using zip ties. It should look just like you see in Figure 7.15. Position the receiver close to the front so it can be close to the speed controllers.

FIGURE 7.15 Zip-tie the receiver to the gondola.

11. Now let's create the connection that will use the servo to move the dowel. Attach a length of timing belt to the dowel, and when it's dry, wrap the belt around the dowel a few times to give it some friction. Figure 7.16 shows how I did it—by stapling the belt to the dowel and then hot-gluing it to heck.

FIGURE 7.16 Staple and glue the timing belt to the wooden dowel and then wrap it around.

12. Once the belt is connected, rotate the dowel so that the propellers are pointing straight down. They should want to do that anyway—thanks, gravity! Then, position the servo arm so it's slanted diagonally forward, as shown in Figure 7.17. Zip-tie the free end of the belt to a #4 screw (1") secured to the end of the servo arm. Thus, when the arm pulls back, the motors will tilt forward. This gives you two directions to propel the blimp: up and forward!

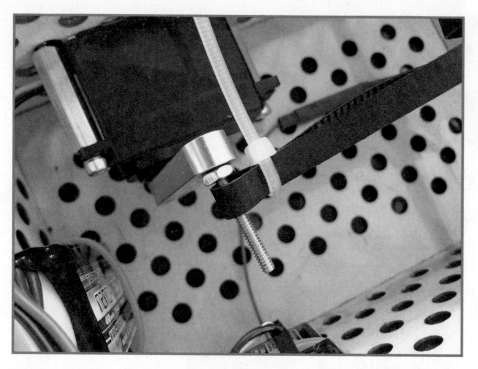

FIGURE 7.17 Zip-tie the timing belt to a #4 screw attached to the servo arm.

13. Wire everything up!

   a. Twist together the red, blue, and black motor wires to the matching wires on the ESC. Figure 7.18 shows how the red one is done.

   If you want to learn an even cooler way to secure wires, I'll show you how bullet connectors work in Chapter 10, "Building a Quadcopter IV: Power Systems." These connectors are absolutely the best way to temporarily connect two wires together.

   b. The other ends of the ESCs have a three-wire plug and separate red and black wires. The red wire from each ESC plugs in to the battery's positive connector, and the two black wires plug in to the battery's ground connector. You will have to twist each pair together into a single wire, as shown in Figure 7.19. Cover up the exposed wire with electrical tape.

   The battery I chose for this project consists of a four-wire charging plug as well as a heavier, main power and ground wires.

   As with the previous step, there is a better way to do this that I detail in Chapter 10. Called a wiring harness, it's a power distribution rig for quadcopters, powering multiple motors with a single battery connection.

FIGURE 7.18 The three motor wires connect to the matching wires on the ESCs.

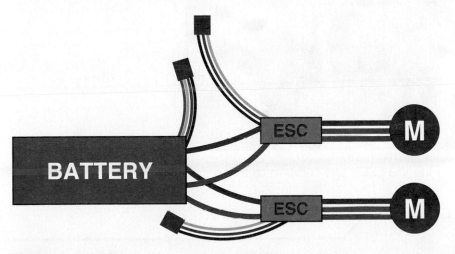

FIGURE 7.19 Connect the leads of both ESCs to the appropriate plugs on the batteries.

    c. The three-wire plugs on the ESCs plug in to the receiver. The black wire goes on the side of the row of pins opposite the antenna, as shown in Figure 7.20. Plug them in to channels 2 and 3. While you're at it, connect the servo plug as well. Put this one into channel 6.

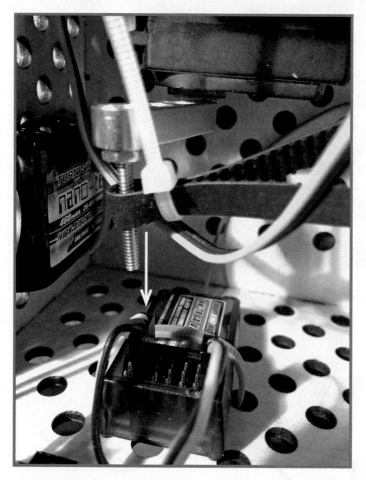

FIGURE 7.20   Plug the ESCs and servo in to the receiver.

TIP

Twisting is actually not a great way to connect wires. In Chapter 8, "Building a Quadcoptor III: Flight Control," I'll explain bullet connectors, industry-standard wiring plugs.

Autonomous Control with an Arduino

Instead of the receiver, we'll add an Arduino to the gondola (see Figure 7.21) as well as a pair of ultrasonic sensors. These electronics will help control the Arduino autonomously.

It's actually very easy to convert the gondola to be run by an Arduino. The ESCs can be pinged by the Arduino's digital pin, effectively playing the same role as the receiver—only with no radio! The instructions come solely from the program. As usual, you will have to program the ESCs first. You can follow your ESCs' directions to program them; otherwise, stay tuned for Chapter 8, in which I delve a little more into the mysteries of ESCs.

Anyway, let's talk sensors. Because it doesn't rely on your eyes to avoid obstacles, the blimp will need its own eyes. We'll install a pair of Ping-compatible ultrasonic sensors—one positioned facing forward to look for obstacles in front of the craft, and the other sensor pointed down so the blimp knows how high up it is. Here are the steps to follow:

1. Use zip ties to attach the ultrasonic sensors (one is visible in Figure 7.21) to the gondola—one pointing down and one pointing forward.

FIGURE 7.21 Ultrasonic sensors are the eyes of the blimp.

2. Yank out the receiver and replace it with the Arduino, wiring it up as you see in Figure 7.22:

   a. Plug the ultrasonic sensors into pins 7 and 8, with their ground wires plugged into GND and their VCC pins connected to the Arduino's 5V supply.

   b. Plug the ESCs' data wires into digital pins 10 and 11 (marked with cyan and pink wires on the wiring diagram).

   c. Plug the servo's data wire into digital pin 9, with its power pin connecting to the Arduino's 5V supply and its ground wire going to GND.

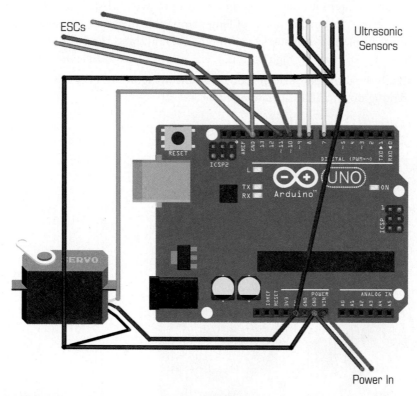

FIGURE 7.22 Swap in an Arduino in place of the receiver.

Code

Upload the following code to your Arduino to control it autonomously. Note that this is not sophisticated code.

```
//This code is based on the PING ultrasonic sensor sketch by David A. Mellis.
#include <Servo.h>
```

```
Servo leftESC, rightESC, axleServo;

const int usPin1 = 7; //belly ultrasonic

const int usPin2 = 8; //forward ultrasonic

long duration1, inches1, cm1;

long duration2, inches2, cm2;

void setup() {

  axleServo.attach(9);

  leftESC.attach(10);

  rightESC.attach(11);

  Serial.begin(9600);

}

void loop()

{

  //let's declare variables and take a reading.

  pinMode(usPin1, OUTPUT);

  digitalWrite(usPin1, LOW);

  delayMicroseconds(2);

  digitalWrite(usPin1, HIGH);

  delayMicroseconds(5);

  digitalWrite(usPin1, LOW);

  pinMode(pingPin1, INPUT);

  duration1 = pulseIn(usPin1, HIGH);

  cm1 = microsecondsToCentimeters(duration1);


  if (cm1 < 2000) //triggers when gondola drops below a 2-meter altitude.

  {

    axleServo.write(100); //turns the props so they're facing down; adjust number as
necessary

    delay(15);

    leftESC.write(100); //adjust amount as necessary for both ESCs.

    rightESC.write(100);

    delay(30);

    axleServo.write(100); //turns the props so they're facing forward again; adjust
number as necessary
```

```
    delay(1000);

}

//let's do the same for the other sensor

pinMode(usPin2, OUTPUT);

digitalWrite(usPin2, LOW);

delayMicroseconds(2);

digitalWrite(usPin2, HIGH);

delayMicroseconds(5);

digitalWrite(usPin2, LOW);

pinMode(usPin2, INPUT);

duration2 = pulseIn(usPin2, HIGH);

cm2 = microsecondsToCentimeters(duration2);

if (cm2 < 3000) //triggers when gondola approaches a wall to within 3 meters.

{

    rightESC.write(100); //adjust number as necessary

    delay(100);

}

}

long microsecondsToCentimeters(long microseconds)

{

  return microseconds / 29 / 2;

}
```

Summary

In this chapter, you immersed yourself in RC and blimps—two areas of interest for drone builders. In Chapter 8, you'll learn more about RC systems, including autopilots and flight controllers. Then you'll build your own!

Building a Quadcopter III: Flight Control

Flight control systems are vital to a successful quadcopter. Simply put, it's difficult to keep a quadcopter in the air if you have to manipulate all four (more, more!) motors manually—not to mention critical and optional features such as being able to manage the power needs of the motors, auto-level the copter, navigate via GPS, and switch back and forth between autopilot and manual control.

This chapter explains how the three parts of the standard quadcopter control rig works: electronic speed controllers (ESCs), flight controllers (FCs), and receivers. After we cover the basics, we will install a commercial Arduino-based autopilot product, the MultiWii, shown in Figure 8.1.

Know Your ESCs

As mentioned previously, electronic speed controllers (ESCs) provide power to the motors so the receiver or flight controller doesn't have to (see Figure 8.2). They typically consist of power input wires, two or three output wires leading to the motor, and a data connection that plugs into the receiver or flight controller. Another feature that some ESCs boast is that they change the battery's DC power into three-phase AC that drives the quadcopter's motors, thus allowing you to run an AC motor using a DC battery.

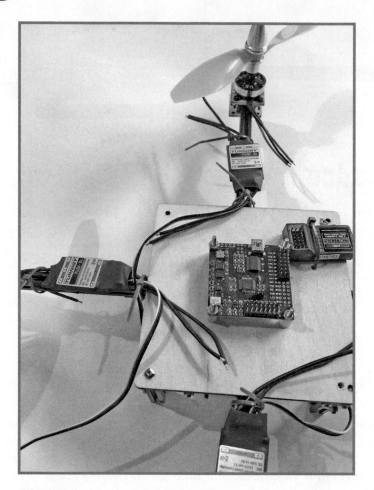

FIGURE 8.1 You install the MultiWii flight controller in this chapter.

ESCs are preconfigured or can be manually configured for a specific type of drone. For instance, an airplane-style UAV is expected to have one or two drive motors and servos for ailerons and the rudder. Some ESCs come preset to cut out the drive power if the battery voltage drops quickly, thus preserving the remaining battery charger for the receiver and control surfaces.

The speed controller market is actually a broad category and has many similar products. It's always a good idea to see what other people with similar projects are doing for speed controllers before you spend money. Fortunately, low-end ESCs don't cost a lot of money.

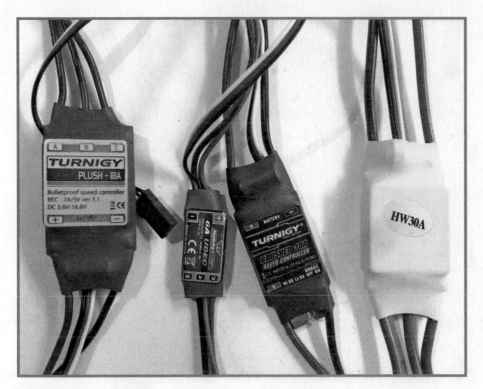

FIGURE 8.2 Choose the right ESC for your project.

Common ESCs

The following are three common but high-quality ESCs:

- **XXD HW30A**—This ESC features a constant current of 30A and a peak rating of 40A for bursts of 10 seconds (see Figure 8.3). It has a safe startup feature where it won't launch instantly if the craft is powered up with the throttle all the way over. You can configure battery type and cutoff settings using your transmitter, as I'll explain later in this chapter.
- **Turnigy Brushed 30A ESC**—This ESC has two motor wires instead of three because it drives a regular, old DC brushed motor rather than an AC outrunner like the motors we use in the quadcopter project (see Figure 8.4). You don't configure this ESC using a transmitter like you do other models. Instead, you use jumpers (little plastic-and-metal conductors) to set just two options: battery type and brake. These are inexpensive ESCs, running around $8 at HobbyKing.com.

FIGURE 8.3 The HW30A is a good all-around 30-amp ESC.

FIGURE 8.4 This 30A ESC is designed for brushed DC motors.

- **HobbyKing 6A UBEC**—This slender little package, shown in Figure 8.5, is also very inexpensive, costing only $7. You can hook it up to two or three LiPo cells, but it's really not made to handle a lot of amperage. It's called a UBEC (universal battery eliminator circuit) because it cuts off the voltage when the amps drop below 0.5.

Programming ESCs

Programmable ESCs use microcontroller chips to store certain settings, such as battery type, the aircraft's configuration (for example, plane vs. copter), braking, throttle range, and many others. The more money you spend on the ESC, the more options it's likely to have.

FIGURE 8.5 This little HobbyKing model is good for low-amp projects.

Simpler ESCs use jumpers (removable plastic-and-metal conductors) to change just a few settings, such as whether the brake is on or what type of battery is being used. More complicated and expensive ESCs may be programmed through the transmitter, with audio tones guiding the user through various menus.

Here is how to configure the Turnigy Plush 30A ESC we're using in the quadcopter project. The ESCs you buy may not have similar options, so be sure to check the data sheet before you purchase them. Lastly, remember that if you use an ESC without the rest of the RC gear, such as by triggering it with an Arduino, you will still need to program it first using the transmitter.

1. Switch on the transmitter, move the throttle stick to the bottom position, as shown in Figure 8.6, and then connect the battery pack to the ESC.

   The ESC should emit a special tone, wait five seconds, and then make another tone. This tells you that program mode has been entered.

   The ESC then cycles through its menu of eight items, each with its own style of beep (such as a single fast beep) that help you recognize which option you're working on. When you reach an option you want, move the stick to the bottom to select that option.

   - **Brake (one fast beep)**—Allows you to set the brake (two fast beeps) or disable it (one).
   - **Battery (two fast beeps)**—One beep indicates a Li-On or Li-Poly battery, whereas two beeps indicates a NiMh or NiCd battery.
   - **Cutoff Mode (three fast beeps)**—Tells the ESC what to do if the voltage drops. One beep tells the ESCs to reduce power, hopefully allowing for a soft landing. Two beeps has the batteries shut off when they're dry.
   - **Cutoff Threshold (four fast beeps)**—Gives you three options for how much power should be kept in reserve when cutoff is triggered. Low is one beep, medium is two beeps, and high is three beeps.

FIGURE 8.6 Use your transmitter to program your ESC.

- **Startup Mode (one long beep)**—Sets how fast the quadcopter takes off. Normal (one beep) makes the drone shoot upward at maximum speed and is recommended for fixed-wing aircraft. Soft (two beeps) lowers the throttle range somewhat so it launches more gently. Super soft (three beeps) offers an even gentler launch. These last two are recommended for quadcopters.
- **Timing (one long and one short beep)**—Defaults to low (one beep) timing, which is fine for most motors. High-efficiency and multi-pole motors sometimes use medium (two beeps) or high (three beeps).
- **Set All to Default (one long and two short beeps)**—Changes all of the previous settings to their default settings.
- **Exit Program Menu (two long beeps)**—Gets you out of the program menu so you can fly!

For those users who prefer a visual cue rather than an audible one, you can often buy inexpensive "programming cards" with LED indicators, so you can see the status of your ESC at a glance. Not only does this method not require a transmitter, it's faster than toggling through a bunch of options with a joystick.

Receiver

The second piece of the flight electronics package is the receiver. As I mentioned in previous chapters, the typical receiver is actually not all that interesting. Receivers have only four areas of differentiation: channels, antenna style, frequency, and modulation. However, choosing between receivers often just means deciding how many channels you want.

Each channel is a separate stream of information. Typically this determines how many separate devices can be controlled by the transmitter. Therefore, a very basic plane's landing gear might be controlled on one channel and its rudder on another. Channels range in number from three to eight or so for normal hobbyist gear; high-end equipment often has well over a dozen channels.

Multicopters are more complicated for the simple fact that they have more motors than many receivers have channels. The flight controller handles the separate motors, and the receiver connects to the FC. Instead of directly controlling the motors, the transmitter works in aeronautical terms—roll, yaw, and pitch—as well as controls the throttle. *Roll* is rotation along the Y axis—think of an arrow rotating as it flies through the air, *Yaw* is rotation along the Z axis, like a spinning top, and *pitch* is rotation on the X axis (doing somersaults). The pilot controls these four factors and learns how to steer by working the yaw.

Because of this simplified way of steering multirotors, you only need four channels to control even an octorotor. If you're flying a quadcopter, springing for a 16-channel receiver that costs $100 is overkill.

Receivers are super simple. They passively wait for the right signal on the right frequency and then send on an instruction to the flight controller or ESCs. The transmitter is where the real excitement can be found.

The bottom line is that you should buy the receiver that goes with your transmitter. If you don't have a transmitter, buy a set that has both the transmitter and receiver. An example I often see cited—and that I also suggest—is the HobbyKing HK-T6A. It's a six-channel, 2.4GHz transmitter and receiver set that costs around $25 (see Figure 8.7). It's just about the cheapest combo available, making it great for hobbyists!

FIGURE 8.7 The HobbyKing receiver I used in this book is fairly typical for a six-channel receiver.

Flight Controller

The flight controller (FC) is the brains of the quadcopter. It usually consists of a microcontroller with various sensors added on, such as an accelerometer, barometer, magnetometer, and so on—basically anything the quadcopter might need for autonomous flight.

However, even the simplest FC helps, even during manually controlled flights. FCs usually auto-level the craft, allowing the operator to focus on steering and not merely keeping the drone in the sky.

A lot of their features are autonomous. For instance, you don't have to do anything with your transmitter to get it to auto-level; it just does so in the background so you can enjoy flying.

FCs can also be set to automatically take certain steps if a malfunction occurs. If the craft detects that it is falling out of the sky, for instance, it could deploy a parachute. (Damage to crashed quadcopters is a sad reality of the hobby.)

Flight Controller Examples

The following sections discuss three FCs, which are just a sampling of the many different products on the market.

Hoverfly Open

The Hoverfly, shown in Figure 8.8, is part of a family of related controllers, ranging from the simpler, entry-level HoverflyOPEN to the HoverflyPRO, which has built-in GPS and costs $900.

FIGURE 8.8 The HoverflyOPEN flight controller connects to ESCs and the receiver.

The HoverflyOPEN can be configured to serve as FC for quadcopters, hexcopters, and octocopters, and it can be mated with any five-channel transmitter/receiver combo.

Hoverfly even offers a mini controller just for the gimbal, a moveable camera mount many quadcopter fans add to their drones.

You can learn more about Hoverfly's products at http://www.hoverflytech.com/controllers/.

Ardupilot

Another example of a fight controller is the Ardupilot, which is an Arduino-controlled autopilot...as you might expect (see Figure 8.9). The platform was created in 2007 by the DIYDrones community, an online forum for drone enthusiasts.

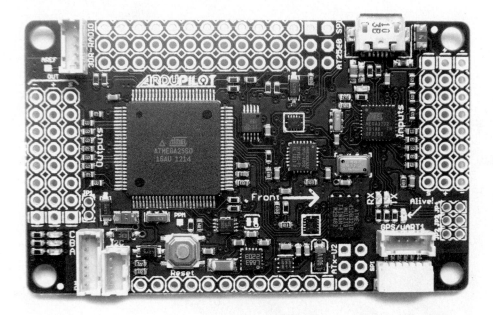

FIGURE 8.9 The Ardupilot combines Open Source Hardware with the excitement of drones! (credit: Explore Labs (Creative Commons))

Earlier versions consisted of Arduino add-on boards called "shields," specialized for connecting servos and ESCs and studded with sensors such as accelerometers and magnetometers. Later on, the Arduino chip was placed directly on the same circuit board as the rest of the components, thus saving space and weight.

One cool aspect of the project, a hallmark of the Open Source Hardware movement, is that anyone may contribute to the Ardupilot project, and several spinoffs have been created, including ArduRover, ArduPlane, and ArduCopter, all specialized boards designed for specific kinds of drones. You can learn more at Ardupilot.com.

MultiWii

The MultiWii is a simple and elegant flight controller originally designed to allow users to steer drones using Wii nunchuks and Motion Plus controllers (see Figure 8.10). Subsequent work has brought it the more mainstream realm of transmitters and receivers, although the Wii software still supports it.

FIGURE 8.10 I used the MultiWii in the quadcopter project.

The MultiWii has an onboard Atmega 328P microcontroller chip, the same as is used in the Arduino UNO. It has up to eight motor outputs and two servo outputs: accelerometer and gyroscope.

Many different flavors of MultiWii boards are sold by as many different stores—as an open-source project, anyone can offer their own flavor of MultiWii-compatible hardware. Learn more about the software initiative at MultiWii.com. You can also buy the same MultiWii board I got. It's available at HobbyKing.com and costs around $30.

Installing the Flight Electronics

Let's continue with our quadcopter project by adding the components already described in this chapter: the ESCs, the receiver, and the controller. We'll tackle them one at a time. However, it won't be until Chapter 10, "Building a Quadcopter IV: Power Systems," that you'll actually wire everything together.

Parts

You'll need the following parts to install the flight electronics. With the exception of the last item, they are all items you can get from any craft store or big-box hardware store.

- Drill and bits
- Velcro bands
- Zip ties
- Double-sided tape
- Four #4 × 1" screws with washers and nuts
- Four #4 × 3/8" standoffs (SparkFun P/N 10461)

Installing the ESCs

Use Velcro bands or zip-ties to secure the parts to the booms, but do so loosely, as shown in Figure 8.11, so you can still remove the ESCs. You will be adding special connectors to all the wires in Chapter 10. Once those connectors are in place, you can pull the ties tight.

Needless to say, position the ESCs so that the three motor wires are closest to the motors, and the power and data wires are toward the center of the copter. You may also want to put the ESCs on the underside of the booms; this keeps the top looking clean and positions the ESCs close to the battery pack, which will get added to the bottom of the wooden square.

Installing the Flight Controller

The MultiWii has four mounting holes on the PCB. Use the holes to mark the wood and then drill holes in the wood corresponding to the ones on the circuit board. Thread four #4 screws with washers through from underneath, and then screw on nylon standoffs. Finally, put the MultiWii on the screws and secure with nuts. It should look like Figure 8.12.

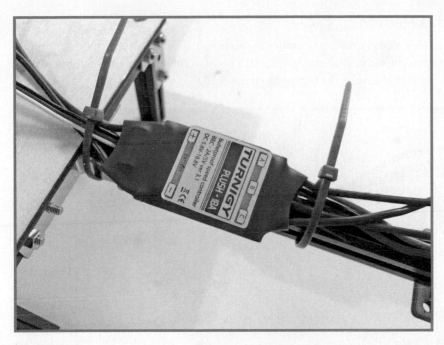

FIGURE 8.11   Don't cinch the zip ties too tightly.

FIGURE 8.12   Drill holes and attach the MultiWii with screws and standoffs.

Installing the Receiver

The receiver used here doesn't have any mounting options except a zip tie, although some double-sided tape might work. Drill the appropriate holes in the wooden plate and then thread a zip tie through the holes, as shown in Figure 8.13.

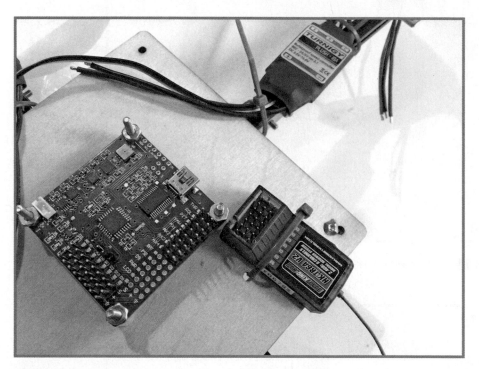

FIGURE 8.13 Thread a zip tie through the holes to secure the receiver.

Summary

This chapter explained the installation of the flight electronics, consisting of the ESCs, the receiver, and the flight controller. You also learned how to customize the settings on the speed controllers to make them work better with your project. In Chapter 9, "Drone Builder's Workbench," you explore a selection of essential drone-building tools you may need.

Drone Builder's Workbench

So far this book has mentioned a lot of tools, both big and small. This chapter covers some of the hardware tools you'll need. I've divided these tools into a number of categories:

- **Design It**—Covers the tools for conceptualizing and designing your drone
- **Drive It**—Describes screwdrivers and other turners of bolts
- **Measure It**—Determine lengths, widths, diameters, and so forth
- **Cut It**—These are tools you need to make cuts and drill holes
- **Wire It**—Tools to help you work with electronics
- **Attach It**—Glue, tape, and other solutions of securing things
- **CNC It**—Discusses Computer Numerically Controlled machines such as laser cutters, 3D printers, mills, and so on

Design It

The first step of any project is to design it. First, you sketch it up on paper; then you move to the computer to do the actual design. At least that's the way it works for me! Here are some of tools you'll need:

- **Pens and pencils**—Even if you always keep your notes on the computer, you should always keep some regular drawing utensils, as shown in Figure 9.1, if only for marking things to be cut.
- **Notebook and graph paper**—You'll need paper to go with those pens. I use a composition notebook, such as the one shown in Figure 9.1, but also will use quad-ruled paper (that is, graph paper).
- **Fritzing**—I love this open-source software, downloadable at Fritzing.org. I use it to do my wiring diagrams (found throughout the book) as well as circuit board design for manufacturing my own PCBs.
- **Inkscape**—Inkscape (Inkscape.org) is like Adobe Illustrator or CorelDraw for people who don't want to (or can't) spend a lot of money. All three packages are vector drawing programs, which means that everything they create is represented as a series of curves, lines, and shapes. Contrast this with a raster art program such as Photoshop, which records its images like a photo. I mostly use Inkscape to create designs for the laser cutter.

■ **SketchUp**—Another popular software tool robot builders use is SketchUp, a 3D modeling program with a vast library of shapes. Whether you want to design a skyscraper in full scale or simply design the broom closet of your dreams, SketchUp is the ticket. It comes in free and commercial versions, with tons of cool features unlocked with the pro version.

FIGURE 9.1 Chances are, designing your next project will involve a sheet of paper and a pencil.

Drive It

Enough design—let's dive in to see some actual hardware tools. This section is devoted to drivers and wrenches for nuts, bolts, and so on. Follow along with the callouts on Figure 9.2.

A. Multitool—Every drone builder, as well as every self-respecting nerd, should have a multitool. I use this sweet SOG Tools No. B61, which includes 22 tools in a very robust (one might say "weighty") package.

B. Needle nose pliers—Great for picking up small things or reaching into the guts of a project.

C. Socket set—The SK socket set (P/N 91848) is very sweet. Keep in mind that you don't need anything this fancy. You won't use it every day at first, but when you start getting into bigger drones with heavier hardware, you'll be grateful to have it.

D. Hex wrenches—A lot of drone hardware uses hex bolts, with sizes ranging from really tiny to fairly big. Accordingly, you'll need a wide range of hex wrenches (also known as "Allen wrenches"). I have a bunch of them, and I'm always finding ones I don't have. Your best bet is to invest in a big set of mixed standard and metric wrenches. You can get a 30-piece set for under $20.

E. Jeweler's screwdrivers—Many drone parts are tiny, so these equally minuscule screwdrivers are a necessity. I use Tekton (P/N 2987), which does the job fairly well. It includes both standard and Phillips drivers, plus hex male and hex female as well.

F. Screwdrivers—I use regular old screwdrivers all the time in my projects. I'm sure you have several already, but you'll want both standard and Phillips, with a variety of sizes.

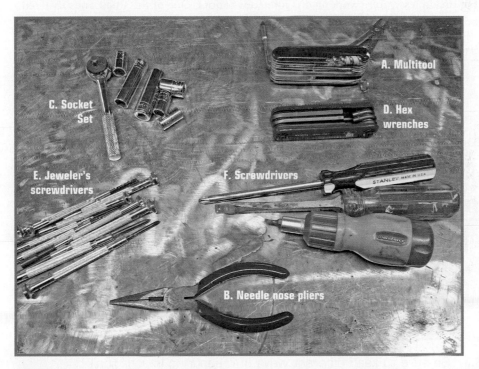

FIGURE 9.2 Drivers are a drone builder's bread and butter.

Measure It

When you're building a kit, you don't have to measure anything, because someone already did it for you. When you're building your own robot, however, you need to be able to measure things accurately. Figure 9.3 shows what I use:

A. Caliper—You use this to measure the diameters or widths of things. It's actually quite inexpensive; you can get a solid no-frills caliper for under $25. I use mine all the time.

B. Ruler—You always need a short ruler. This one from Adafruit (P/N 1554) is cool for electronics nerds because it has circuit board features such as standard via diameters, trace widths, and the footprints for all the usual surface-mount electronic packages.

C. Protractor—Just the thing for measuring angles. You can find cheap ones with the school supplies, or nicer ones at hardware stores.

D. Measuring tape—A measuring tape is one of those must-have tools. It doesn't have to be a big one unless you're working on eight-foot robots. One of the small ones that sell for under $5 will do just fine.

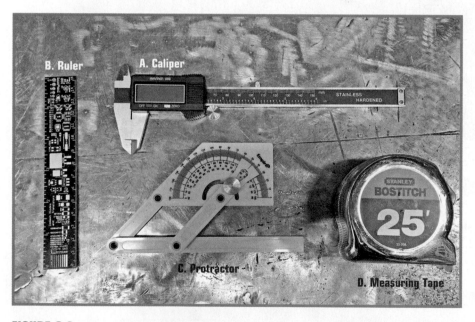

FIGURE 9.3 Be sure to keep different ways of measuring in your toolkit.

Cut It

Even when you're dealing with kits, you'll often have to cut things. Having a variety of tools, as shown in Figure 9.4, will help your cutting activities go better.

A. **Drill**—This DeWalt is my number-one homeowner's tool, and I find myself using it for robotics as well. It is used mostly for drilling small holes, but also for driving wood screws—because who wants to turn a screwdriver?

B. **Dremel**—Every hobbyist's favorite rotary tool, this cordless Dremel (P/N 8220) is great for cutting, polishing, drilling, and so forth.

C. **Hacksaw**—When you're dealing with metal, you'll always need a hacksaw.

Although not shown in Figure 9.4, X-Acto knives are another tool you'll need. These blades are great for delicate model work, such as cutting a fragile plastic connector or detaching a laser-cut balsa part.

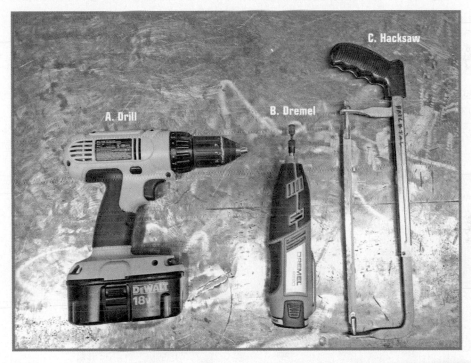

FIGURE 9.4 Being able to cut through wood, plastic, and metal is a necessity.

Wire It

Next up are a number of items you might need for your electronics work (see Figure 9.5). Because I already covered soldering equipment in Chapter 7, "Blimp Drone Project," I include only a token mention here.

A. Power supply—When one is powering a prototype, the temptation is to connect the battery pack likely to be employed when the project is done. However, debugging while batteries drain can be problematic. A power supply delivers an unending feed of DC electricity, with the user able to adjust the volts and amps to fit the needs of the project. Although not a must for beginners, having a power supply is definitely convenient. I swear by this Extech 382202, which sells for around $100.

B. Wire cutter—You'll be working with wire, and this Vice Grip wire cutter/stripper (P/N 2078309) is a great all-around tool to have.

C. Automatic wire stripper—I love this no-name wire stripper. You just put a wire into the jaws and squeeze the tool shut, and it automatically strips the insulation off of the end.

D. Soldering equipment—I already covered soldering in Chapter 7, so this iron is a stand-in for all the stuff I recommended in that section.

E. Multimeter—This handy tool measures volts, resistance, conductivity, and other aspects of the electronics world. I suggest the BenchPro BP-1562, sold by Jameco Electronics. It's a great basic meter that will get you started for around $10. The Fluke meter pictured in the figure is much nicer and runs around $80.

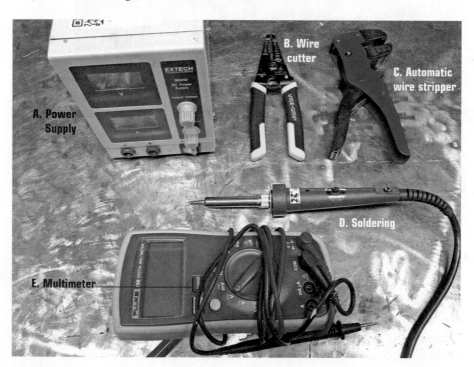

FIGURE 9.5 Working in electronics? Buy these tools.

Attach It

It is every drone builder's secret sorrow that nuts and bolts do not always work. Sometimes you must use hot glue, double-sided tape, or zip ties to secure the parts of the drone. Embrace this fact with these great options, shown in Figure 9.6:

A. **Double-sided tape**—I use this stuff a lot. Always get the foam type (I swear by Scotch's 1" tape) and not the type that looks like clear tape with sticky on both sides. The foam tape has peel-off backing that doesn't get messed up.

B. **Super glue**—I use one-shot mini containers, because it's foolish to hope to reseal and reuse a tube of super glue. Chances are it'll be dried up or glued shut.

C. **Zip ties**—Everyone's favorite attach-everything tool, the zip tie. You can get great multipacks of ties at any hardware store.

D. **Velcro**—A great way to manage wires. It accomplishes the same as zip ties, but may be a little classier.

E. **Hot glue**—Everyone's gotta have a hot glue gun.

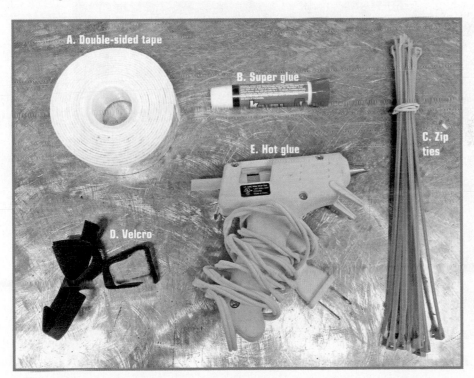

FIGURE 9.6 Need to attach something and screws won't work? Your solution may be one of these tools.

CNC It

Lastly, I want to present a layperson's guide to a number of complicated and expensive items that I consider to be extremely useful to drone builders. Although a lot of these machines are very expensive—upwards of $10,000!—there are more budget-friendly models, as well as alternatives to ownership. In the latter case, I'm talking about tool libraries, community colleges, and hackerspaces (organizations that are eager to have you borrow instead of buy). Here are some larger categories of CNC tools:

■ **Laser cutter**—This tool (seen in Figure 9.7) is used to cut through plastic and wood, following a design you have created in a vector art program such as Inkscape. This is great for drone makers because it allows you to learn from others' explorations. Want a laser-cut drone chassis? You can find a bunch of designs online, already created and available for download.

Lasers cut cleanly and precisely, allowing you to piece together complicated structures without a lot of glue or other connectors.

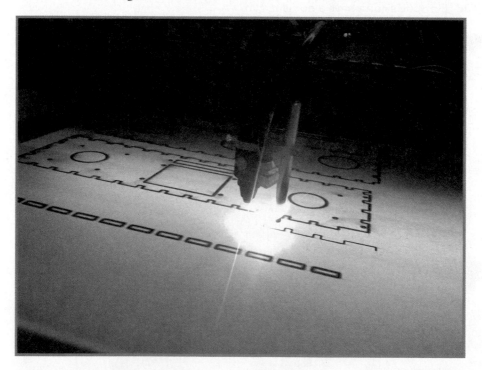

FIGURE 9.7 The best way to cut wood is with light.

■ **3D printer**—A 3D printer (seen in Figure 9.8) lays down melted plastic filament, building up three-dimensional objects, layer by layer. As with laser patterns, 3D printer designs abound on the Internet, allowing you to riff off of other makers' projects. It's a great way to learn and gets you to a prototype faster than making everything by hand.

One of the best uses of a 3D printer in the drone world involves printing up connector parts, allowing you to mate dissimilar components together. I give an example of this in Chapter 12, "Building a Quadcopter V: Accessories," where I print a camera mount as well as a base intended for a different model of camera.

FIGURE 9.8 A 3D printer outputs 3D shapes in plastic.

- **CNC mill**—This is a generic term for a computer-controlled router or rotary tool that carves a shape out of wood, metal, or plastic. It's another must-have tool that unfortunately can set you back quite a bit of money.

  A CNC mill looks kind of like a Dremel or similar rotary tool on computer-controlled gantries (see Figure 9.9). Unlike the laser, which cuts with light, the CNC mill uses router bits (they look kind of like drill bits) and these have a tendency to wear out.

FIGURE 9.9 A CNC router is a very cool, but very expensive tool.

Summary

This chapter presented a ton of tools that a drone builder might reasonably need in the course of his or her experiments. In Chapter 10, "Building a Quadcopter IV: Power Systems," you'll add a battery pack to your quadcopter and also learn how to wire everything up.

Building a Quadcopter IV: Power Systems

The quadcopter is starting to come together, with flight electronics installed in Chapter 8, "Building a Quadcopter III: Flight Control." In this chapter, you'll learn about various types of batteries, as well as how to install one in the quadcopter. You'll also learn about bullet connectors, the drone world's favored way of attaching components to each other. Finally, you'll build and install a wiring harness (as shown in Figure 10.1).

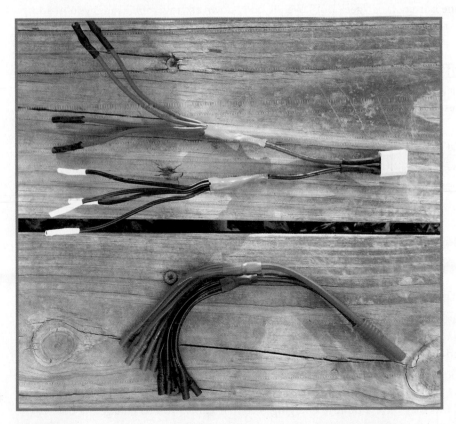

FIGURE 10.1   You'll assemble a wiring harness in this chapter.

Choosing a Battery

Let's get started choosing your battery. However, in order to make a good decision, you'll need to know your options. Here are four criteria to help you decide on which one to buy:

- **Volts**—Unlike regular old AA cells, rechargeable lithium and nickel batteries have differing voltage ratings. Make sure to check the rating before you buy, or you may find the voltage too much or too little for your project.
- **Milli-amp hours (mAh)**—This is the total theoretical power output by the battery. The more mAh, the longer the battery lasts.
- **"C" rating**—This is the maximum safe discharge rate of the battery. This number is multiplied by the mAh to give you the maximum number of amps the battery can safely discharge. For instance, a 460 mAh battery with a C rating of 25–40 can thus discharge between 11.5A and 18.4A. This is important because it tells you how powerful the propellers will be. Note that the low number is the continuous discharge, and the high number is the max.
- **Energy density**—This is how many amps you get for the size. This is kind of an obscure stat, but one battery fans "geek out" over. It boils down to how much mass you're adding to your aircraft for the results you get.

Battery Types

Typically, drone builders use two battery types to power their creations: nickel batteries and lithium batteries. Let's go over them one at a time.

Nickel Batteries

Nickel-metal hydride (NiMH) cells are the rechargeable batteries that put old standby NiCad (nickel–cadmium) out to pasture (see Figure 10.2). They have the advantage of being commonplace—you can buy NiMH batteries in most convenience stores.

The downside is that they have the AA form factor, which requires a battery holder and means more weight. It is also more of a hassle to recharge the cells. What's more, they don't have the energy density of the Lithium batteries, packing only 140–300 Wh/L.

Lithium Batteries

Most of the batteries used in drone operation are lithium batteries—either lithium polymer (LiPo) or lithium ion polymer (Li-ion). These batteries, shown in Figure 10.3, have superior specs compared with nickel batteries, outputting 3.7V versus 1.2V for NiMH.

The configurations for LiPo and Li-ion batteries are much more conducive to being installed on a drone. Not only are they typically contained in a plastic wrap and thus require no battery holder, but they have a built-in charging wire that allows you to charge the battery without removing it from the drone.

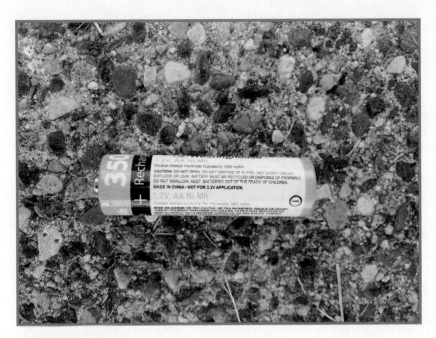

FIGURE 10.2 NiMH batteries are a popular choice for drones.

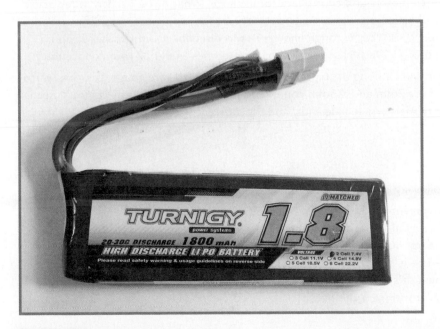

FIGURE 10.3 LiPo batteries are the go-to power solution for most drones.

Although very cool batteries, LiPos are potentially hazardous. However, following a small number of safety rules will see you through:

- Don't short-circuit or immerse the battery. This can potentially cause a fire.
- Don't puncture the LiPo cells. This can also cause a fire.
- If the battery catches fire, use sand to smother it. Lithium takes the oxygen out of the water and keeps on burning!
- If the battery starts to swell up, discontinue use.
- Use a commercial LiPo charger to recharge the batteries. I suggest the Turnigy C3; it's an inexpensive charger that recharges all two- and three-cell LiPos.
- Never mix and match battery types. Use only batteries of the same chemistry together.

Adding Bullet Connectors

Now that you have selected a battery, let's start wiring up the drone. First, however, let's begin with a lesson on bullet connectors, which provide a typical way of connecting components in the RC and drone world.

Why Bullet Connectors?

How do you connect all the quadcopter's components? Surely you wouldn't solder in connections between motors and controllers, for instance. What if you needed to replace one or the other? Bullet connectors are a popular choice for joining together different systems (see Figure 10.4). They're tough enough not to disconnect without catastrophic damage to the copter, in which case the least of your worries would be your connections!

The main differentiation among the types is connector size, with wire gauge dictating which size to purchase. I use 2mm and 3.5mm connectors in the quadcopter project. You'll need an equal number of male and female ends, and the different sizes do not mate with each other.

Parts List

You'll need the following parts and tools to add bullet connectors:

- **Soldering equipment**—In Chapter 7, "Blimp Drone Project," I showed you how to solder; the chapter also lists the equipment you'll need.
- **Bullet connectors**—You can find these at any RC store.
- **Heat shrink**—This is special nonconductive rubber tubing that shrinks down to coat your component. It's like a smarter version of electrical tape. Adafruit sells a multipack (P/N 344). Otherwise, you can find it in any electrical supply or hardware store.

FIGURE 10.4 Bullet connectors are the preferred connection method in the RC world.

Steps for Adding Bullet Connectors

Bullet connectors are metal inserts and sockets that are soldered onto the ends of wires. Once they're all soldered up, the ends snap together securely, and you're in business. Here are the steps for adding your own bullet connectors:

1. **Choose genders.** Plan out in advance how you want to arrange everything so you don't have any rude surprises. One common way to organize the genders is to start with the motors and give them male connectors. Obviously, the motor side of the ESCs would have to have female connectors. The power supply wires on the other side of the ESCs are given male connectors. Figure 10.5 shows a male and a female bullet connector.

2. **Strip the wires.** Strip a little bit of insulation off of the end of the wires, as shown in Figure 10.6, where you can see the wires of four ESCs getting prepped simultaneously using a piece of wood with holes laser cut in it. Many ESCs and other components come with their wires pre-stripped.

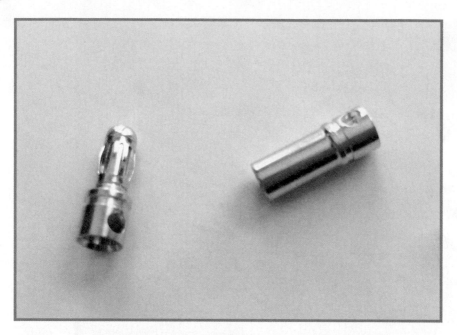

FIGURE 10.5   A male (left) and a female (right) bullet connector.

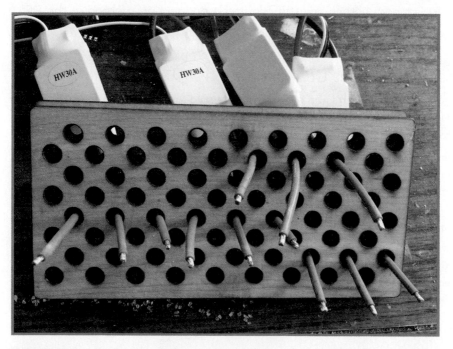

FIGURE 10.6   Strip a little insulation off the ends of the wires.

3. **Tin the wire ends.** Tinning means to coat with solder. This helps the component stick to another tinned component. First, twist up the wire strands (if any) and use a soldering iron to heat up the wire end. Then coat it in solder, as shown in Figure 10.7. Do this for every wire that will be taking a bullet connector. Note that some components, such as the Turing Plush ESCs I use in this project, come pre-tinned as well as pre-stripped.

FIGURE 10.7 Tin every wire that will have a bullet connector.

4. **Fill the connectors.** The equivalent of tinning for the bullet connectors is to fill them up with a glob of solder, as shown in Figure 10.8. There is a hole in the side of the connector. Fill up the solder to this level.

FIGURE 10.8   Fill up the bullet connectors with solder.

5. **Solder them on.** Put a tinned wire end into one of the connectors. Put your soldering iron tip to the hole, which will help melt the solder inside it. You may want to add some more hot solder around the top of the connector if it seems loose. Figure 10.9 shows how it should look.

FIGURE 10.9   Solder on the bullet connector.

6. **Add the heat shrink.** Put about an inch of heat shrink on the end of each wire, making sure the contact portion of the male connector is left free. Use your soldering iron barrel to warm the heat shrink so it coats the base of the connector as well as the end of the wire. Figure 10.10 shows successful male and female connectors.

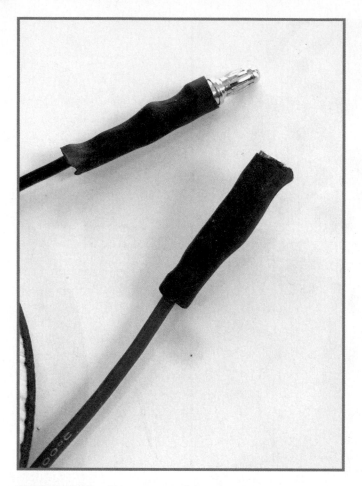

FIGURE 10.10 Put some tubing on the end and shrink it down!

7. **Connect!** Finish by snapping the motor's male plugs into the as ESCs' female plugs, as shown in Figure 10.11. You'll be connecting the actual battery in the next section. While you're at it, you'll want to cinch the zip ties tight to keep the ESCs in place.

FIGURE 10.11   Connect the ESCs to the motors.

Assembling the Wiring Harness

Adding the battery probably seems tricky, if not downright intimidating, for the simple fact that you need to figure out how to connect the power and ground leads of four ESCs to a single battery.

This is typically handled one of two ways. First, many drone builders buy or design a power distribution board, which is a fancy way of saying a circuit board preconfigured to combine four (or more) power and grounds into one pair. Look for them at a hobby store; they typically cost in the range of $5 to $10 on the low end.

Many drone builders instead choose to solder up their own wiring harness. This is composed of four or more ground wires twisted and soldered into a single connector, and an equal number of lead wires similarly configured. Figure 10.12 shows an example of a wiring harness. We'll take this approach for the quadcopter project.

FIGURE 10.12 A wiring harness splits the power lines into one for each component requiring power.

Parts

You'll need the following parts and tools:

- Soldering equipment.
- Wires: 12 gauge and 16 gauge stranded, black and red for each.
- Heat shrink. Sparkfun P/N 09353 offers a nice assortment.
- XT60 connector (). This is a great connector I describe later.
- Bullet connectors. They come in different sizes, but I use 3.5mm.

Steps for Assembling the Wiring Harness

Let's get started on the wiring harness. Follow along with these steps:

1. **Cut 16-gauge wires approximately five inches in length.** You'll want one red and one black wire for each component to be powered—certainly one per motor! For each wire, strip off a half-inch of insulation from one end and a quarter-inch from the other end. Figure 10.13 shows how it should look.

FIGURE 10.13 You'll need a black and red wire for each component.

2. **Solder on bullet connectors.** Put a female bullet connectors on the end of each wire that has the quarter-inch of insulation removed. Add heat shrink as you did before, as shown in Figure 10.14. These connectors will plug into the ESCs' power wires, which have male connectors.

FIGURE 10.14   Add bullet connectors.

3. **Cut the 12-gauge wires.** Next, cut two five-inch lengths of heavier 12-gauge wires. For each wire, remove about half an inch of insulation from one end and a quarter-inch from the other end. Figure 10.15 shows how it should look.

4. **Solder the wires together.** Tin the exposed strands of the 16-gauge and 12-gauge wires (the ones with half-an-inch of insulation removed). Next, lay the 16-gauge wires alongside the 12-gauge wire, as shown in Figure 10.16, and solder them together. Cover with heat shrink.

FIGURE 10.15 You'll need a big power wire and a big ground wire.

FIGURE 10.16 Solder together the wires.

5. **Add the XT60.** The XT60 is composed of a pair of bullet connectors with a robust housing that not only shields the exposed metal from short circuits, but also prevents the battery from being plugged in backwards, which is likely to damage the drone's

electronic components and possibly start a fire. You'll need to solder the female connector to the battery's leads and the male connectors to the 12-gauge wires. Treat them just like bullet connectors in terms of how you solder them, and don't forget to add the heat shrink. Figure 10.17 shows how it should look.

FIGURE 10.17 Add XT60 connectors to the battery ends and the 12-gauge wires.

6. **Attach everything!** Zip-tie the battery to the wooden platform. However, don't connect the XT60 ends together—we're not ready to fly yet! Grab the ESC and motor combos (see Figure 10.18) then connect the 16-gauge ends of the wiring harness to the ESCs by snapping together the bullet connectors.

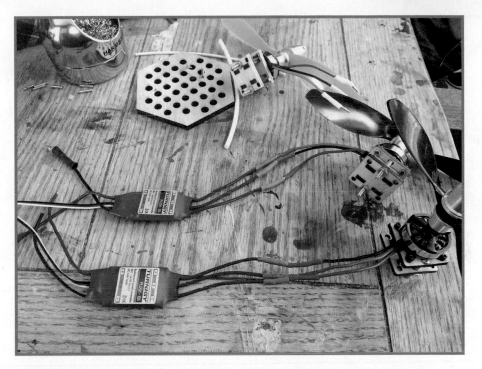

FIGURE 10.18 Snap the bullet connectors together to connect the drone's motors and ESCs to the wiring harness.

Wiring the Flight Controller and Receiver

The final series of steps involve plugging the ESC control wires and the receiver into the FC:

1. Use female-to-female servo extension wires (such as SparkFun P/N 8738) to connect the flight controller to the receiver, like so:

   - Connect the pins marked THR (throttle) to Channel 1 on the receiver. Make sure you plug in the wires correctly: Ground should be toward the edge of the receiver, and toward the edge of the MultiWii, as shown in Figure 10.19.

   - Connect the pins marked ROL (roll) to Channel 2.

   - Connect PIT (pitch) to Channel 3 on the receiver.

   - Connect YAW (yes, it's yaw) to Channel 4.

FIGURE 10.19 Wire the flight controller to the receiver.

2. Connect the ESCs' triple wires to the FC. Plug the wires into the pins marked D9, D10, D3, and D11 on the MultiWii, making sure to keep the black wire toward the edge of the flight controller's PCB. It should look like Figure 10.20.

FIGURE 10.20 Connect the ESCs to the flight controller.

Summary

In this chapter, you learned about the two basic categories of power cells that drone builders typically employ in their projects. Then you soldered up the battery's wiring harness and installed it. Finally, you wired up the flight controller and receiver. In Chapter 11, "Waterborne Drone Project," you'll create a boat drone to explore your local pond or kiddy pool.

Waterborne Drone Project

In this book you've learned about seemingly every kind of drone imaginable, but there are still more types to discover! In this chapter, you build the waterborne drone, a floating robot with soda-bottle pontoons, which is shown in Figure 11.1. Before you get to the project, however, you'll need to understand the advantages and disadvantages of building water-based drones. You'll also learn about two important topics: how to waterproof your electronics, and how to set up an XBee mesh network, which is another way to control a drone. You'll use this know-how to build a remotely controlled boat steered by a cool little handheld controller you'll assemble yourself.

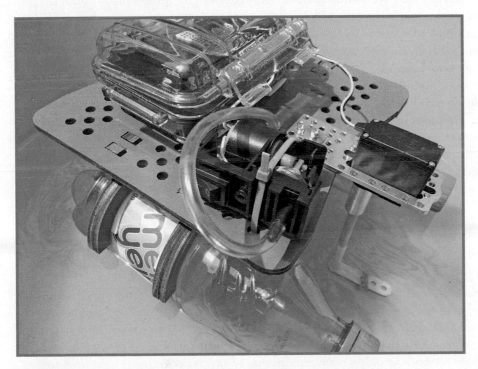

FIGURE 11.1 The Soda Bottle Boat explores a body of water.

Realities of Waterborne Electronics

Everyone knows if you drop a cellphone in the bathtub, it will get fried and probably won't work again. We get that water and electronics don't get along. However, there are more considerations than just waterproofing to keep in mind while building such a craft.

Disadvantages of Waterborne Electronics

Let's go over the hazards and inconveniences of operating a waterborne drone:

- **Water damages electronics**—In one sense, this well-known hazard is just as bad as everyone supposes. An Arduino or other electronic module may be destroyed by immersion, leaving your drone dead in the water. On the other hand, you shouldn't feel like everything is DOA when immersed. Ordinary brushed DC motors, for instance, are perfectly happy operating underwater, as long as you allow them to dry out. Don't put them in salt water, however, because it will quickly corrode the motor's innards.
- **Displacement of water is required**—Dead electronics and corroded motors aren't the only hazard. Let's face it, the drone has to stay afloat, either with pontoons or some sort of boat shape that displaces water. The latter works only as long as it doesn't flip over! That said, there are a lot of options. The boat in Figure 11.2 is made out of a brownie pan!
- **Space is required**—Unlike quadcopters and rovers, boat drones can't be played with in your backyard, unless you have a big pool. Going down to the neighborhood park is always an option, as long as there aren't any kids in the wading pool. Ironically, if you are around a body of water larger than a lake, it may very well be too rough and choppy to safely operate a small craft.
- **Drones disappear or get easily damaged**—There is a real possibility of never seeing your drone again if you send it out onto a large body of water. If your ROV sinks into the silt at the bottom of a river, you're unlikely to recover it unless you have a strong cable attached to it. By contrast, quadcopters—which do occasionally fly off into never-never land—mostly can be recovered when they fail, if only in pieces!

Advantages of Waterborne Electronics

Don't let all that doom and gloom get you down. There are also some very cool things about this type of drone:

- **Less friction**—Waterborne drones require less force to propel than other types, because friction atop the water is much less than, for instance, on the ground (for ground-based drones). Sailboats illustrate this principle very well. Even a tiny breeze will get a sailboat moving when it's on the water. This works to your advantage because you can use relatively crazy methods to propel your craft. In that spirit, this chapter's project uses an air pump to push the craft along.

- **Simple design**—Waterborne drones typically are simple vehicles, usually requiring only two motors—one to propel the craft and one to steer. Quadcopters obviously need four, but many types of aerial drones have more—six or eight is not uncommon. Even the ground-based rover you'll build in Chapter 13, "Making a Rover," uses four motors.
- **Multiple configurations**—You can choose among many different configurations, ranging from submersibles to hovercraft to surface boats. There isn't just one kind of floating drone to make. Have fun and be creative!

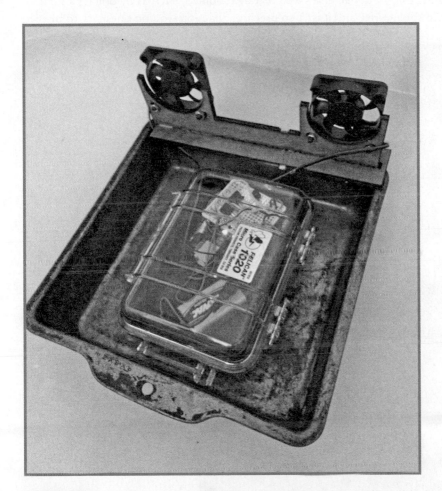

FIGURE 11.2 You can make anything into a boat, including this brownie pan.

Waterproofing Your Electronics

The major bugaboo of making floating drones, as everyone knows, is getting water on your electronics, an accident very likely to short-circuit and destroy the components. The best way to protect your electronics is to have them sealed inside a waterproof enclosure. The following sections describe three different routes for enclosing the electronics. (There is also a fourth option—a chemical treatment—that is even more different.)

Sandwich Container

When in doubt, go low rent! The Rubbermaid seven-cup container pictured in Figure 11.3 is relatively cheap and easily obtainable. It's also readily modified, such as by drilling a hole in it to accommodate some wires, and it isn't so precious that you wouldn't hesitate to alter it.

FIGURE 11.3 This food storage container makes an excellent project enclosure for not a lot of money.

Another advantage to going this route: You can get the things in any number of sizes, ranging from "smaller than your palm" all the way up to "can hold Thanksgiving dinner" in size. I mentioned sandwich-sized in the heading mostly because it's a good configuration for an Arduino and battery pack.

Note that spending a lot of money in this category won't necessarily get you a tighter seal, which is mostly what you want. You may find the dollar store box is just as good as Tupperware.

Pelican 1000-Series

These tough-as-nails enclosures are designed to hold cell phones and other valuable electronics, and basically protect them from pretty much everything, including immersion (up to 1 meter for 30 minutes), sudden movement, and crushing. They have a huge rubber gasket inside that seals the interior against moisture but also cushions the contents against impacts.

That may seem a little radical, but the price is right: The Model 1010 (left) and Model 1020 (right) cases pictured in Figure 11.4 cost only $9 and $15, respectively, varying by store. If you don't like those sizes, Pelican offers a variety of cases, ranging up to a giant trunk that takes two people to lug. As with the Rubbermaid tub, you're likely to need to modify the cases to allow wires to pass in and out. I buy my Pelican cases off Amazon, but other online retailers offer them as well.

FIGURE 11.4 Pelican's 1000-series cases are very small but waterproof and tough.

Sealing a Tube

A more do-it-yourself approach is used by the OpenROV submersible, pictured in Figure 11.5. The designers sealed the batteries inside plastic tubes, which feature end-caps sealed against moisture.

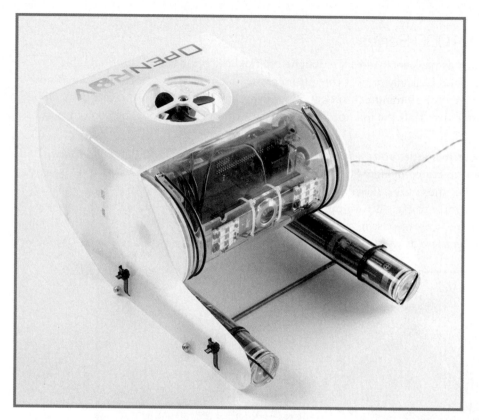

FIGURE 11.5   This OpenROV submersible seals its batteries inside watertight tubes.

Other versions of the ROV use PVC piping. You can buy PVC in any hardware store, and it's most frequently associated with plumbing. It consists of all sorts of watertight fittings such as curves and forks. Many tinkerers make furniture out of PVC, for its structural uses, but not many have taken advantage of its watertight nature. This stuff was meant to keep water in; it's no wonder it also can keep it out.

XBee Mesh Networking

Let's shift gears a bit and talk about another way to control your drone. More specifically, let's discuss how to use the technology that will control your waterborne drone.

Mesh networking using an XBee wireless module (see Figure 11.6) makes for a flexible way to link multiple Arduinos together. In this chapter's project, you'll have just two nodes: the drone and the controller. However, the technology can connect many more nodes than that—256 or more, depending on the model of radio.

FIGURE 11.6 XBee radios offer simple off-the-shelf wireless control.

A mesh network differs from more hierarchal networks in that all of the nodes are equal: When you send a command, every radio in the net hears it. In order to command one or another radio, you have to specify in the software to pay attention only to commands meant for that radio. This isn't your only option: XBees come in multiple configurations, and you can specify different types of networks.

XBee radios are based on ZigBee, an industry standard wireless protocol with many different flavors and spinoffs. Entry-level XBees are rated at 1 milliwatt, and have a range of 80 feet indoors and 300 feet outdoors. A "Pro" version, which costs more, offers a correspondingly better range: 140 feet indoors and a whopping 4,000 feet outdoors. Although the Pro version impresses, even the basic model is great for short-range drones.

For more information on setting up your own XBee network, check out Bildr's XBee tutorial (http://bildr.org/?s=xbee) and adafruit.com has a great one as well: https://learn.adafruit.com/xbee-radios/overview.

Project: Soda Bottle Boat

Now that you're up to date on waterproofing techniques and XBee networks, let's tackle the project for this chapter: the floating drone, consisting of a wooden platform atop a pair of soda bottles (see Figure 11.7). It's propelled by an air pump and is controlled by a wireless remote control you'll build yourself. Let's begin!

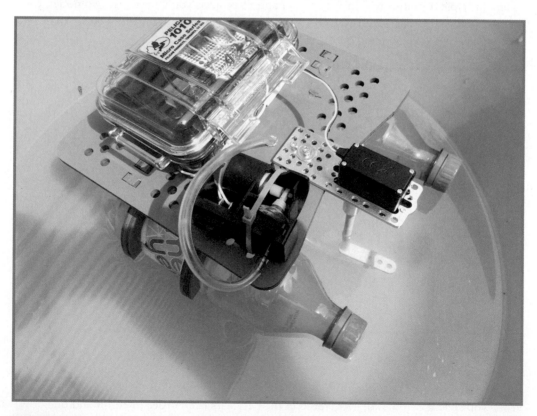

FIGURE 11.7 The Soda Water Boat propels itself with an air pump and can be controlled wirelessly.

Parts

You'll need the following parts to build your Soda Bottle Boat:

- Laser-cut chassis. You can download the pattern at http://www.thingiverse.com/jwb. Cut the chassis out of 1/8" (3mm) plywood.
- Two soda bottles. I used typical 20oz Mello Yello bottles, 2.7" in diameter across the label, a bit wider above the label, then narrowing to a 1" diameter neck. If your bottles are different, you'll want to adjust the laser pattern accordingly.
- Pelican case (P/N 1010).
- Arduino UNO.
- Two XBee radios. I suggest the Series 1 from SparkFun (P/N 8665). You'll need two!
- Two XBee breakout boards to manage the radios (SparkFun P/N 11373).
- Battery-powered air pump. The lighter the better! You can find these suckers cheap at any pet store. I yanked the pump out of a Marina (P/N 11134).
- Tubing. I used Tygon B-44-3 beverage tubing, but your tubing doesn't have to be food-safe. Anything with a 1/4" outer diameter and a 3/16" inner diameter would work well. I bought it off Amazon.
- Servo. A waterproof one like the Hitec P/N 35646S is great, but way overpowered for your needs. Even a sub-micro servo would be adequate for this job. I ended up using a Hitec HS422. You can buy these and other servos at ServoCity.com.
- Servo plate. I used a servo plate (Actobotics P/N 575144) to secure the motor.
- Shaft adapter for the servo (Actobotics P/N HSA250). This secures the end of the dowel while connecting securely to the servo's hub.
- Dowel, 0.25" diameter, 3–4" in length.
- Three buttons. You'll want "momentary" buttons that release when you lift your finger. SparkFun P/N 9190 is a good one.
- Two proto boards. I suggest either "perforated bare phenolithic prototyping boards" from Jameco (P/N 616690) or the SparkFun ProtoShield (P/N 7914).
- Male header pins (SparkFun P/N 12693).
- TIP120 Darlington Transistor (Adafruit P/N 976). This electronic switch triggers the pump when the Arduino gives it a signal.
- Two LEDs. We'll use just a couple of them—any old LED will do.
- Two resistors, 220 ohm. SparkFun sells an assortment (P/N 10969) that includes some 220s.
- A 1N40001 diode (Adafruit P/N 755).
- Double-sided tape.
- Wire.
- Zip ties.

Building the Drone

Once you have all your parts together, it's time to build your boat. After you're done with the boat, you'll build the controller.

1. Laser out the chassis. I show the bottle rests getting glued in Figure 11.8. To be honest, making a chassis is something you can do with masking tape and cardboard; it doesn't have to be as fancy as this. As long as it holds the bottles in check and keeps the enclosure out of the water, it will be fine.

FIGURE 11.8 The chassis design includes these bottle rests.

2. Assemble the chassis by gluing the parts together, as you see in Figure 11.9. Once the glue is dry, cover it with a couple coats of your favorite spray paint to protect it against moisture.

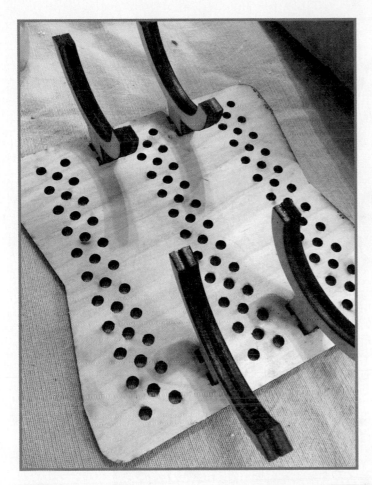

FIGURE 11.9 Assemble the chassis.

3. Attach the Pelican case (shown in Figure 11.10) using double-sided tape. Alternatively, you can use zip ties. Even more radical, you can drill mounting holes in the base and screw the chassis into place, although this will make it less watertight, obviously. I have the chassis resting on the soda bottles but they haven't been attached yet.

4. Solder up the breakout board of the drone's Arduino by following along with these sub-steps:

    a. Add the header pins to the breakout board, as seen in Figure 11.11; they should conform to the pins on the Arduino. You may even want to use the Arduino itself to help keep the pins in place while you solder them in from above, just to keep everything straight.

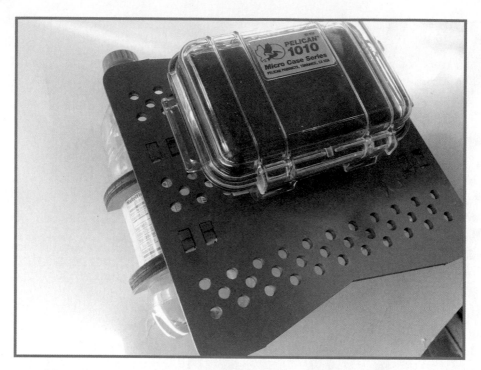

FIGURE 11.10 Attach the Pelican case to the chassis.

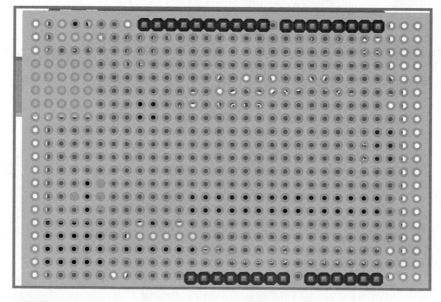

FIGURE 11.11 Solder the header pins into the board.

b. Solder the XBee Explorer to the board and use wires to connect its pins to the appropriate Arduino pins. Data out (the pink wire in Figure 11.12) goes to pin 3 of the Arduino, while data in (blue) connects to pin 2. Power (red) and ground (brown) connect to their respective pins on the Arduino.

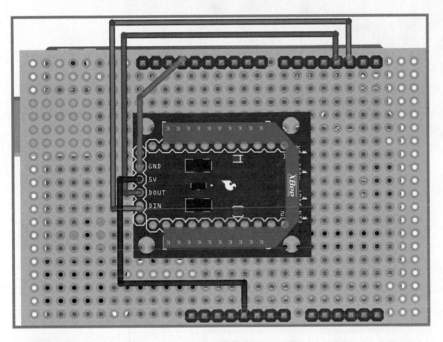

FIGURE 11.12   Add the XBee Explorer.

c. We're going to add a status LED that lights up when the Arduino has power. Connect 5V to the positive lead of the LED with a 220-ohm resistor in between. I use an orange wire in Figure 11.13. Connect the negative lead of the LED to ground, shown as a black wire in the figure.

d. Add the TIP120 to the board, with the leftmost pin (called the base) connected to Arduino pin 11 via a 2.2K resistor. This is shown as a purple wire in Figure 11.14. The rightmost pin, called the emitter, connects to the Arduino's ground pin, and this is depicted as a cyan wire in the figure.

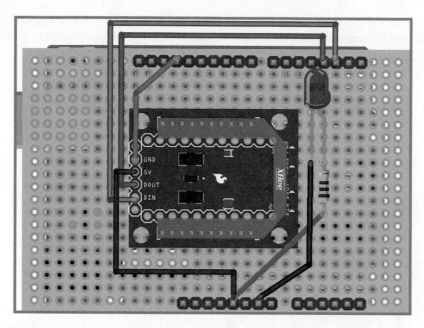

FIGURE 11.13 Solder in a test LED and resistor.

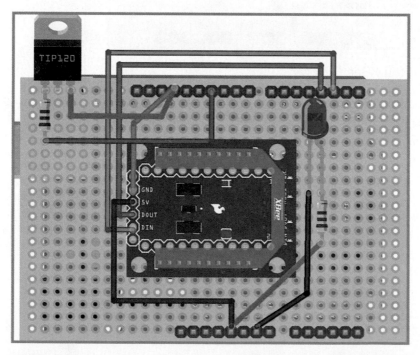

FIGURE 11.14 Solder the TIP120 transistor into the board.

e. Add the pump and diode. The diode helps keep the motor from feeding back on itself, and it's a good idea to include one every time you use a DC motor, of which this pump is an example. One lead (shown as a green wire in Figure 11.15) of the motor connects to the 3.3V pin of the Arduino, while the other one (yellow) plugs into the center pin of the transistor, called the collector.

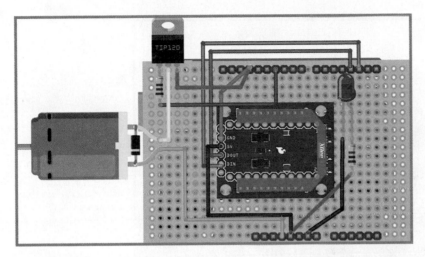

FIGURE 11.15 Connect the pump and diode.

f. Add the servo. The yellow and white striped wire (seen in Figure 11.16) is data and connects to pin 9 of the Arduino. The other two wires, red and black (each with stripes), connect to 5V and ground, respectively. You'll usually encounter servos with plugs on the end of the wire. If this is the case, you can solder in some male header pins and simply plug in the servo.

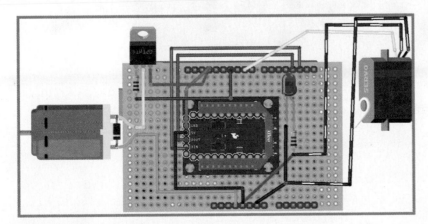

FIGURE 11.16 Next, attach the servo.

g. When you're ready to power up the drone, connect a 9V battery to the battery holder and plug it into the barrel jack on the Arduino.

5. Attach the air pump to the chassis using a zip tie (see Figure 11.17). The output tube can point in any direction, as a length of plastic tubing will direct the airflow in the direction we want.

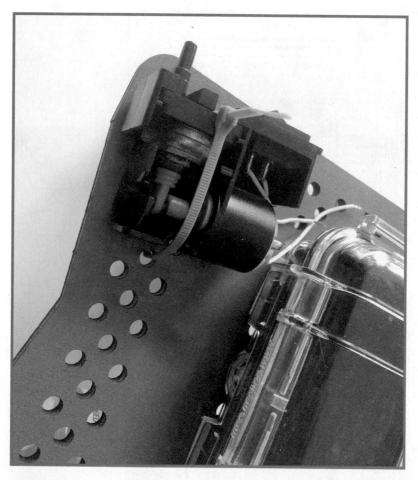

FIGURE 11.17 Attach the air pump with zip ties.

6. Attach the servo to the chassis with the help of the servo plate. It should look like Figure 11.18.

FIGURE 11.18 Attach the servo to the chassis.

7. Attach the dowel to the servo using the shaft coupler, as shown in Figure 11.19. Just slide it onto the splines of the servo's hub, then secure with the servo-horn screw sold with the motor. Slide the dowel into the 5mm end of the coupler and secure it using the accompanying set screw. While you're at it, glue the tube holder onto the end of the dowel—you printed this small piece out with the rest of the chassis.

8. Put the electronics package in the Pelican case, making sure you have enough wire to allow the pump and servo to be installed in their proper places (see Figure 11.20). You may need to alter the box to allow the wires to exit; although this will affect the water tightness of the box, it's a necessary sacrifice!

FIGURE 11.19 Mount the dowel to the servo.

FIGURE 11.20 Place the electronics in the Pelican case.

9. Plug a tube into the pump's output and run it through a chassis hole—drill one as needed. Figure 11.21 shows how it should look.

FIGURE 11.21   Slide the tube through a hole in the chassis.

10. Zip-tie the tube to the tube holder on the dowel, as seen in Figure 11.22.

FIGURE 11.22   Angle the output tube to propel the drone forward.

11. In the final step for the drone's hardware, attach the soda bottles to the chassis using double-sided tape, as seen in Figure 11.23. Simply run the tape along the inside of the bottle rests.

Building the Controller

The controller is similar to the drone's electronics package in that it has an XBee and Arduino, but substitutes a couple of buttons for the motors. Follow along with these steps:

1. Assemble the electronics package. Follow these sub-steps to complete the assembly:

   a. Add the header pins to the breakout board, as seen in Figure 11.24, just as you did with the drone's board.

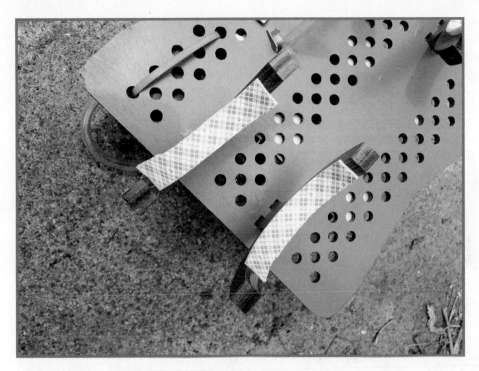

FIGURE 11.23   Add double-sided tape to the bottle rests.

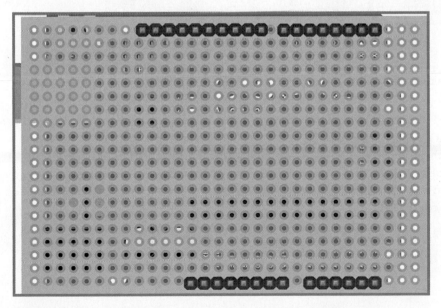

FIGURE 11.24   Solder the header pins into the breadboard.

b. Solder the XBee Explorer to the board and use wires to connect its pins to the appropriate Arduino pins, just as you did before. I used the same wire colors as before: Data out (the pink wire in Figure 11.25) goes to pin 3 of the Arduino, while data in (blue) connects to pin 2. Power (red) and ground (brown) connect to their respective pins on the Arduino.

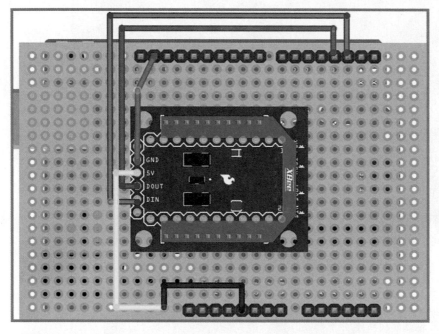

FIGURE 11.25 Add the XBee Explorer.

c. Solder in the status LED as you did before: Connect 5V to the positive lead of the LED with a 220-ohm resistor in between. I use an orange wire in Figure 11.26. Connect the negative lead of the LED to ground, shown as a black wire in the figure.

d. Add a pair of buttons. They each have 5V going into one lead (purple wires in Figure 11.27); the negative lead goes to a digital pin on the Arduino—pins 6 (orange wire) and 7 (green) for the two of them. Finally, connect the negative pin to ground with 2K resistors in between. This is depicted using black wires in the figure. You're done with the controller!

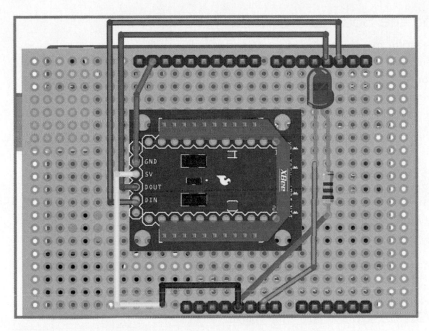

FIGURE 11.26 Solder in a test LED and resistor.

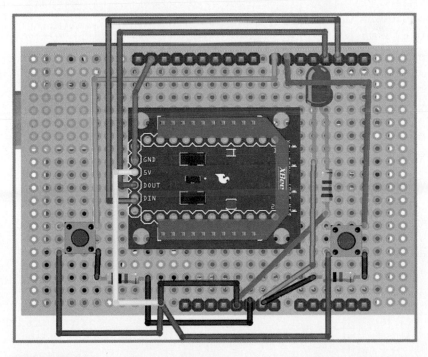

FIGURE 11.27 The buttons control the direction the boat travels.

Programming the Soda Bottle Boat

Fascinatingly, at least for me, the nature of the XBee network is such that the controller and the drone are equals—theoretically, the drone could send a command to the controller. Because of this, you can literally run the same code in both units. This is what we'll do. So, upload the following code to both Arduinos:

```
#include <Wire.h>
#include <Servo.h>
//lets initialize a number of variables and constants
Servo myservo;
const int pumpPin = 11;
const int button1Pin = 6;
const int button1Pin = 7;
int button1State = 0;
int button2State = 0;

void setup()
{
  myservo.attach(9);  // attaches the servo on pin 9 to
                         the servo object
  pinMode(11, OUTPUT);
  pinMode(6, INPUT);
  pinMode(7, INPUT);
  pinMode(button1Pin, INPUT_PULLUP);
  pinMode(button2Pin, INPUT_PULLUP);
}
void process_incoming_command(char cmd)
{
  switch (cmd) {
  case '1': //left
    myservo.write(30); //turn the servo to 30 degrees
    delay(15);
    digitalWrite(pumpPin, HIGH);
    delay(1000); //turn on the pump for 1 second
    digitalWrite(pumpPin, LOW);
    break;
  case '2': //right
    myservo.write(150); //turn the servo to 150 degrees
    delay(15);
    digitalWrite(pumpPin, HIGH);
    delay(1000);
    digitalWrite(pumpPin, LOW);
    break;
```

```
  case '3': //forward
    myservo.write(90); //turn the servo 90 degrees
    delay(15);
    digitalWrite(pumpPin, HIGH);
    delay(1000);
    digitalWrite(pumpPin, LOW);
    break;
  default: //if no command is issued, the pump stops
    delay(1000);
    break;
  }
}


void loop()
{
  if (Serial.available() >= 2)
  {
    char start = Serial.read();
    if (start != '*')
    {
      return;
    }
    char cmd = Serial.read();
    process_incoming_command(cmd);
  }
  // Let's interpret the button-presses.
  button1State - digitalRead(button1Pin);
  button2State - digitalRead(button2Pin);

  if (button1State == HIGH) && if (button2State == LOW) {
    Serial.write('*');
    Serial.write(1);
  }
  if (button1State == LOW) && if (button2State == HIGH) {
    Serial.write('*');
    Serial.write(2);
  }
  if (button1State -- HIGH) && if (button2State == HIGH) {
    Serial.write('*');
    Serial.write(3);
  }

}
```

Summary

In this chapter you learned about waterborne drones and actually building one. In Chapter 12, "Building a Quadcopter V: Accessories," we'll finish off the hardware build of our quadcopter with a 3D-printed camera mount that allows you to take pictures in the air.

Building a Quadcopter V: Accessories

So far in the quadcopter project, I've covered such important topics as flight control, motors, and batteries. Now for the fun category: everything else! It turns out there is a bunch of extras you can purchase or build for your bird. In this chapter, you'll learn about some of the options available for accessorizing your drone. Then, you'll add a protective plate and a camera mount to the quadcopter—you can see the completed drone in Figure 12.1.

FIGURE 12.1 You'll finalize the quadcopter's physical build this chapter.

Add Accessories to Your Quadcopter

You've got a quadcopter—now what? Many quad pilots seek to add accessories such as cameras, first-person video rigs, and protective plates. Let's go over just a handful of the numerous options out there.

Camera

The most flashy accessory you're likely to encounter, the camera, has an obvious appeal: You can take pictures from a vantage point impossible for humans to attain. In fact, it's probably the most popular accessory, and some people even make money by selling drone footage, though recent laws have clouded this matter somewhat.

One subcategory of mount is the motorized, controllable turret called a gimbal. A gimbal enables you to control the shooting angle of the camera, in addition to triggering the camera's shutter. Most of the time gimbals are used with quadcopters, but not always! Fixed-wing planes and rovers also benefit from having these devices installed.

Many aerial photographers choose the GoPro Hero (GoPro.com), which sets the bar for durable, waterproof (to 40m), high-speed photography. You'll find a large GoPro fan base with countless techniques, 3D-printed mounts, and third-party add-ons to share. Another popular model is the Contour ROAM, shown in Figure 12.2. It's cheaper than a GoPro but has fewer features. Later in this chapter, I demonstrate adding a similar camera to the quadcopter.

FIGURE 12.2 The Contour ROAM is a common choice for drone photography.

First-Person Video (FPV)

A hypothetical onboard GoPro might shoot some great video, but it suffers if there is no one to look through the viewfinder! Sometimes you want to see what the camera will be shooting, and that's where FPV comes in. An FPV setup consists of a small, low-resolution camera, with a transmitter and antenna. The video feed is viewed on a matching receiver, which includes a small monitor. Intriguingly, some FPV suites include video goggles, thus allowing you to see through the quadcopter's eyes!

One of the intriguing aspects of FPV usage is the nascent phenomenon of drone sports. I'm talking racing, with pilots guiding their craft through courses while peering through their drones' eyes. One company, Game of Drones (gameofdrones.com), organizes "aerial combat" events where the last quadcopter left flying wins the day. They even sell tough "Hiro" airframe kits for participants or anyone else to build.

Figure 12.3 shows a close-up of drone-builder Steve Lodefink's quadcopter, showing off its FPV camera in front. You can also get a good look at the protective dome covering the electronics.

FIGURE 12.3 This quadcopter's FPV lens sticks out through the protective plastic dome (credit: Steve Lodefink).

Landing Gear

Landing gear legs not only keep the drone off the ground, but also make room for a camera mounted on the bottom, thus protecting it from impacts. You can buy a kit such as the robust (and big!) combo package pictured in Figure 12.4. It includes mounting pads so you can attach a camera underneath the copter that's protected by the landing gear legs. There are many kits out there, usually specialized for a particular model of quadcopter. However, in many instances they can be adapted to another type of drone. The kit in the figure is a no-name clone designed for the DJI F450 quadcopter, but it can be used in a similarly sized quadcopter with a center-plate connection system.

FIGURE 12.4 Many quadcopter pilots swap in different landing gear.

Of course, I always suggest drone builders check out Thingiverse and similar 3D-printer design sites. They offer tons of different printable legs, including many (as with the commercial kits) intended for specific models of drones. In fact, the parts I designed for this book's quadcopter project are available on my Thingiverse page, http://www.thingiverse.com/jwb.

Parachute

The dread of every quadcopter pilot is to face the sight of the bird falling out of the sky and smashing into pieces on the ground. Gravity can be so unreasonable! One solution is to add a parachute that deploys automatically if power is lost.

Parachute systems definitely cannot be described as being mainstream in the quadcopter world, but a few commercial systems do exist. Skycat Recovery Launchers (Skycat.pro) cost $600, which sounds like a lot, but considering you might have an expensive drone with an even more expensive camera on it, $600 seems reasonable. The Skycat recovery system is shown in Figure 12.5.

FIGURE 12.5 The Skycat recovery system deploys a parachute when the quadcopter loses power (credit: Skycat).

The use of onboard parachutes may also be the law someday, as increased use of quadcopters can lead to more folks bonked on the head by falling drones.

Protective Plate or Dome

You sometimes see quadcopters with a protective plate or dome guarding the electronics. Copters crash, and you wouldn't want to see your lovely microcontroller smashed to pieces. It's one thing to crack your propeller, but quite another to see your $200 flight control package kicking up a divot.

Obviously, if your drone hits a rock or concrete slab, no plate in the world is going to save it. However, the many casual and relatively gentle mini-crashes can also do damage to your quadcopter.

When building your own drone, you might simply want to create your own plate, but commercial options are also available. Many of these ship with the airframe you buy, but you can also find standalone products that can be adapted to fit any quadcopter.

One intriguing angle involves repurposing an old camera dome. I'm talking about those half-sphere globes that keep moisture and inquisitive fingers off of the lens. The one shown in Figure 12.6 is too small for our uses, but would be great for protecting an FPV camera, for instance.

FIGURE 12.6 A protective dome can save your electronics package from a hit.

Prop Guards

Prop guards protect the propellers from casual damage, although they're likely to remain the most vulnerable component on a drone no matter what. Most of the time they consist of a simple plastic ring positioned along the prop edge, though some have a more complicated structure, looking like little cages around the props. Still others incorporate chassis elements to protect the propellers, in effect embedding the props inside the chassis.

Because propellers are apt to break, prop guards are available in countless configurations and styles. They can be found in hobby stores and on websites, and even more can be downloaded from Thingiverse, which enables you to print them out. The guard pictured in Figure 12.7 (http://www.thingiverse.com/thing:652455) is designed for the DJI Phantom 2 Vision.

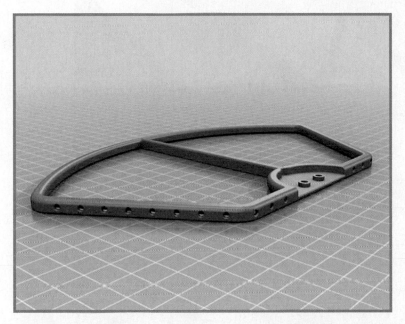

FIGURE 12.7 A rendering of a 3D-printable prop guard (credit: yuppchukno, Creative Commons).

Project: Adding Accessories to the Quadcopter

In the final step of the quadcopter's physical build, you'll add a protective plate, landing gear, and a camera mount to the quadcopter. It's the finishing touches that will make your drone work even better. Figure 12.8 shows the drone with these parts installed.

FIGURE 12.8 The quadcopter project sporting its cool new accessories.

Install the Camera Mount

Two Thingiverse files are used to install the camera mount: renelm's ContourHD mount with GoPro adapter (Thing #423077) and ark19's GoPro Arca Mount V2 (Thing #234654). The beauty of Thingiverse and other collaborative platforms is that these two makers can contribute to your project and not even realize it! You just print their completely compatible designs and connect them. You can see the two parts in Figure 12.9.

FIGURE 12.9 The camera and its freshly-printed mount.

If you don't have access to a 3D printer, you can buy any number of commercial mount options. I consistently recommend going with HitCase (hitcase.com), which specializes in waterproof and impact-resistant phone cases and robust mounting solutions.

1. Print off the aforementioned two Thingiverse pieces, the ContourHD mount and GoPro adapter (http://www.thingiverse.com/thing:423077_) and the GoPro Arca Mount (http://www.thingiverse.com/thing:234654). You can see one of the prints in Figure 12.10.

FIGURE 12.10 The camera mount, hot off the printer.

2. Attach the parts together and secure them with a #4×1" screw. Figure 12.11 shows how it should look.

3. Flip the quadcopter over and attach the camera mount to the underside of the wooden platform, drilling holes as needed (see Figure 12.12). Use #4×3/4" screws.

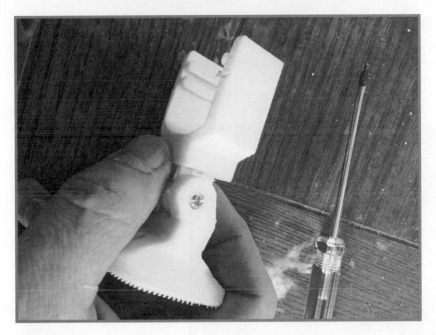

FIGURE 12.11 The two parts get attached together.

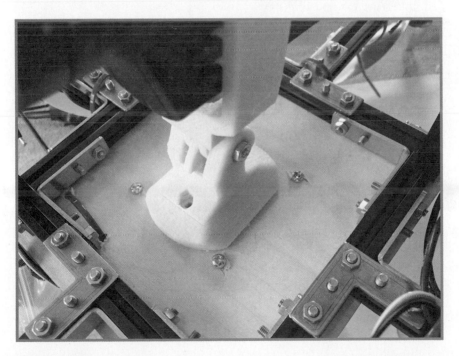

FIGURE 12.12 Attach the mount to the underside of the platform.

Install the Landing Gear

Next, you need to add onto the quadcopter's existing legs to give it proper landing gear. I designed the landing gear in SketchUp, and it isn't pretty, it is functional. The legs have a hollow interior matching the cross-section of the MakerBeam beams. You can see the rendering in Figure 12.13.

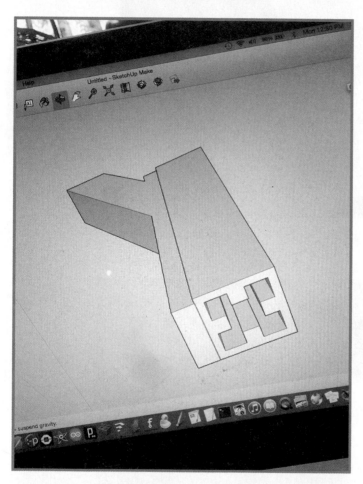

FIGURE 12.13 The landing gear takes shape in SketchUp.

1. Print off the landing gear. The file is available on http://www.thingiverse.com/jwb. Figure 12.14 shows one of the legs printed off.

FIGURE 12.14 The landing gear consists of four of these legs.

2. Once the legs are cleaned up (the inside might require a skinny file!), you'll be able to slide the 3D-printed part onto the MakerBeam legs, as shown in Figure 12.15. For me, the tolerances were so tight I could barely get them on, but if you find yours a little loose, use a hot glue gun to keep them from falling off. If all goes well, you should have a modest amount of clearance for the camera.

FIGURE 12.15 The 3D-printed legs slide on.

Install the Top Plate

The top plate consists of a laser-cut wooden plate with four aluminum standoffs giving it a little clearance for the electronics. Figure 12.16 shows a top view of the quadcopter, showing off the plate.

FIGURE 12.16 The quadcopter's top plate is the final component to add.

1. Laser out the design, which can be downloaded from my Thingiverse page. I used 1/4" birch plywood, my favored laser-cutter fodder. Figure 12.17 shows the plate.

2. Install the plate using four 2" #4 M-M aluminum standoffs or something similar. You can find standoffs in any respectable fastener store, and many hardware stores as well. Figure 12.18 shows the finished quadcopter with the accessories in place.

FIGURE 12.17 The laser-cut plate will protect the top of the quadcopter.

FIGURE 12.18 You're done with the quadcopter's hardware build.

Summary

In this chapter we completed the hardware portion of the quadcopter project by adding a 3D-printed camera mount, and you learned about a number of quadcopter accessories along the way. All we have left is the software! First, however, let's make another drone. In Chapter 13, "Making a Rover," you'll learn how to modify a commercially available car drone to carry an RFID sensor.

13

Making a Rover

The final type of drone you're going to explore in the book is the rover, a car drone that drives around your backyard (or wherever) on missions of exploration and pet-scaring. The RFID-Navigating Nomad, shown in Figure 13.1, consists of a sturdy aluminum chassis equipped with motors, big knobby tires, and a plastic enclosure. Added on to this platform is a sensor package consisting of an RFID (radio frequency identification) sensor and an ultrasonic sensor.

FIGURE 13.1 The RFID-Navigating Nomad steers as directed by RFID tags.

Before we get to the project, however, it's back-to-school time. First, you need to learn about what makes rovers' ground focus both a challenge and an opportunity. Then you'll investigate a number of different chassis options for creating a beginning robot. Finally, you'll learn about how RFID tags are used for navigation by rovers. This is a fun robot!

Advantages and Disadvantages of Rovers

Just like other categories of drones, rovers present certain advantages and disadvantages you'll want to consider when planning your project.

Advantages of Rovers

Here are some of the advantages of rovers:

- You can't lose them! For anyone who saw their waterborne drone sink into the bottom of a lake or gazed mournfully as their quadcopter flew off into the distance, the rovers' inability to leave solid ground comes as a relief.
- Because rovers rely on friction and gravity to stay put, consequently they don't need much power to operate. Depending on their battery capacity, a rover can theoretically stay autonomous for days. If it has a solar array supplying it, it can last even longer. What other type of drone can do that?
- The first two items reinforce this final point: Rovers are the best at being autonomous because they can be left alone for longer periods of time. Imagine a weather station robot, creeping through the bushes and taking sensor readings. You could theoretically leave it there for days, where a quadcopter can stay in the air only minutes.

Disadvantages of Rovers

While reveling in rovers' cool features, you don't want to forget that they aren't perfect. Here are some of the disadvantages of the platform:

- They're a tiny bit boring. Think of it this way: Part of the appeal of making a quadcopter or boat is being able to traverse an unfamiliar environment. Sure, we may not be able to fly like a bird, but we can put a chopper in the air. Being able to roll around on the ground doesn't quite compare.
- As with waterborne drones, there is the problem of where to use a rover. Unless you have a big backyard, you are faced with the challenge of finding a spot that a rover can be used without alarming folks unaccustomed to seeing drones in the underbrush. It's all well and good to have your rover creeping through the bushes, but can you imagine the hubbub if someone got scared? On the other hand, if you have a big yard, you're golden!

Chassis Options

You can make anything relatively solid into a chassis. With rovers, which don't have as much of a weight restriction as quadcopters, you can use steel, wood, plastic, or whatever you want. The primary criterion is whether you can bolt components onto it. For this book I'm presenting a number of options: building a chassis out of a kit, manufacturing one with the use of a 3D printer or other computer-controlled tool, and buying a pre-made robotics platform. Let's go over a few of the possibilities.

3D Printed

Thingiverse and other 3D printing sites feature countless variants of the 3D-printable robot chassis, such as the one shown in Figure 13.2. It's a sendup of the Mars rovers and was designed by Thingiverse user SSG1712. You can find it at http://www.thingiverse.com/thing:835053.

FIGURE 13.2   Like this chassis? Print it up yourself (credit: SSG1712).

Most of these printable frames boil down to some sort of plate with numerous mounting holes on it, typically conforming to mechanical attachments the creators used themselves. Usually the part number of the correct hardware is provided along with the design pattern, theoretically saving you a bunch of time. This sounds like a dream proposition, but you'll encounter a couple of disadvantages. First, 3D printing takes a long time, and it could conceivably take hours to print off a chassis. Second, many makers consider 3D-printed parts to be less durable than a milled or lasered equivalent.

Tamiya

Model-making company Tamiya builds complicated gearboxes and drive trains out of wood, plastic, and rubber. The model pictured in Figure 13.3 includes two DC motors, allowing a microcontroller to steer the assembly by reversing one track or the other. Tamiya sells many of their products as just the chassis, assuming you'll add your robot onto their gearbox. If tinkering with mechanical systems doesn't interest you, this is not a bad option.

FIGURE 13.3 Tamiya: inexpensive, creative, and fun to put together.

mBot

Another angle in the pre-made chassis department, the mBot (http://mblock.cc/mbot/, seen in Figure 13.4) is basically an entire robot than can be programmed as an Arduino or using the Scratch visual programming language (https://scratch.mit.edu/). You'll also notice numerous unpopulated mounting holes on the mBot's chassis, allowing you to add onto the robot. It's like taking the Tamiya drivetrain and adding sensors, buttons, and LEDs to the mix.

FIGURE 13.4 The mBot robot comes equipped with a bevy of sensors.

Arduino Robot

Arduino Robot (shown in Figure 13.5) offers the same idea as the mBot—a pre-built educational robot—except taken to absurd levels, with a built-in LCD, buzzer, audio sensor, button array, and prototyping area, along with a million other features. The downside of Arduino Robot may be its fragility—the entire robot is exposed circuit board material. That and the low wheelbase make it useless for any terrain more challenging than a low-pile carpet. It's also expensive, costing nearly $300—but you get everything!

FIGURE 13.5 The Arduino robot has everything you could ever want, plus space to add on components you never knew you needed (credit: Arduino.cc).

Actobotics Bogie

The mini rover shown in Figure 13.6 (ServoCity.com, P/N 637162) boasts six individually motorized wheels and a rocker-bogie suspension, allowing it to navigate all sorts of obstacles. The bed and legs are formed out of durable plastic, and the suspension system gives it the ability to traverse rough terrain. You'll still need to add a power supply, microcontroller, RC receiver, and other parts to make it into a full-fledged robot.

FIGURE 13.6 The Bogie offers a tough-as-nails platform for an all-terrain rover.

Navigating with Radio Frequency Identification Tags

RFID tags, as they're called (sometimes pronounced "arfids"), don't need power supplies: They are energized by the radio waves of the reader. Tags come in a variety of configurations, as shown in Figure 13.7, including credit cards, keychain fobs, adhesive stickers, and even spikes.

While nearby a powered-up reader, the coil in the tag gets just enough electricity to activate a mini radio, which transmits the tag's unique code. You can get a visual idea of how this works in Figure 13.8.

FIGURE 13.7 RFID tags come in different shapes and sizes.

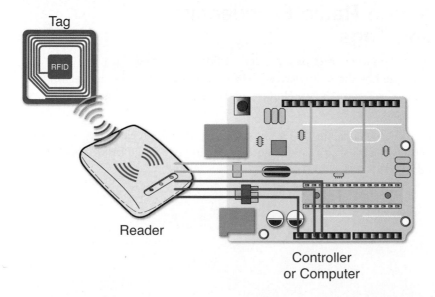

FIGURE 13.8 RFID tags transmit radio waves without their own power source.

The code consists of 32 bits of data, each being a 0 or 1. This translates as a 16-digit string of characters, of which 12 are the actual code. When you take out the start and end markers, you're left with a 10-digit alphanumeric code.

For instance, there might be a code for "turn 90 degrees to the right and then move 10 feet," and another code might be "back up 3 feet." Of course, you'll need to know the codes in advance so you can add them to the Arduino sketch. Another intriguing option involves using the tags on the fly to navigate. Quite literally you could steer the robot by sprinkling tags along its path.

Project: RFID-Navigating Rover

Let's dive in and build the RFID-Navigating Rover, seen in Figure 13.9. It consists of a chassis built from a kit: the Actobotics Nomad. You'll enhance this base with an Arduino and motor shield, as well as an ultrasonic sensor for distance measurements as well as the expected RFID reader. Let's do it!

FIGURE 13.9 The RFID-Navigating Rover uses an ultrasonic sensor to gauge distances.

Parts

Going with a chassis kit reduces the total number of things you'll have to buy, but you'll still need a fair number of parts, including the following:

- Actobotics Nomad rover chassis. This commercially available chassis (SparkFun P/N 13141) forms the backbone of the RFID-Navigating Rover.
- A 3" length of Acrobatics channel (SparkFun P/N 12498).
- Servo with 180% or more degrees of rotation. I used a Hitec HS-422HD (SparkFun P/N 11884).
- Servo mount. I used an Actobotics Servo Plate B (SparkFun P/N 12444).
- Tall servo hub (SparkFun P/N 12227).
- Some 24-gauge stranded wire (Jameco P/N 2187876).
- RFID sensor (SparkFun P/N 11827). This sensor is rated for 125KHz, and you'll need tags to match.
- RFID tags (SparkFun P/N 8310). These are the size of a business card and are encoded with a unique 32-digit ID code that can't be changed. The tag transmits this code when activated.
- RFID breakout board (SparkFun P/N 13030). You do not absolutely have to have this; it is recommended because it breaks out the sensor's 2mm-spaced connectors into more breadboard-friendly 0.1" spacing.
- Ultrasonic sensor. You can buy a Sain Smart HC-SR04 online. A similar sensor is the Makeblock Me-Ultrasonic Sensor (P/N 11001) found at www.makeblock.cc.
- Extension jumpers (Adafruit P/N 826)
- #4-24×1/4" Phillips panhead self-drilling screws, such as Fastenal P/N 0143528
- 9V battery plug. Adafruit P/N 80 does the trick.
- 8×AA battery pack for the motors (Adafruit P/N 449).

Steps

Let's assemble the RFID-Navigating Rover, beginning with building the Actobotics Nomad kit. Note that Acrobatics has a YouTube video describing the build instructions of the chassis, and you can find it here: https://www.youtube.com/watch?v=FAPDkyeAek8. In the meantime, here are the steps to follow:

1. Assemble the ABS chassis. The chassis, shown in Figure 13.10, screws together with the help of the accompanying #6-32 socket-head screws and matching connector blocks. It's called the ABS chassis because it's built out of heavy and durable plastic, called ABS, or acrylonitrile butadiene styrene; it's the same stuff used in LEGO bricks. In addition to its durability, ABS is easily drilled and the chassis comes equipped with mounting holes, allowing you to install any type of hardware you could think of.

FIGURE 13.10   Assemble the ABS chassis.

2. Attach three 4.5" channel lengths to the chassis. Actobotics' key component is the aluminum channel, a beam of metal pierced with dozens of mounting holes. In this step you add three lengths (pictured in Figure 13.11) with the help of more mounting blocks.

   These channels don't actually provide any structural support to the BS—they're there as a platform for sensors and other components. Later on in this build you'll have the opportunity to do just that—add a rotating ultrasonic sensor to the top channel and the RFID sensor to the front channel.

FIGURE 13.11   These channels will eventually hold sensors.

3. Add the 6" channel length to the bottom of the ABS chassis with the help of more connector blocks, as shown in Figure 13.12. You'll also add two different mounts to either side. One mount is a quad pillow block (callout "A" in the figure) with a built-in bearing, allowing a quarter-inch axle to rotate freely. The other attachment is a quad-hub mount (callout "B"), similar to the other one but lacking a bearing.

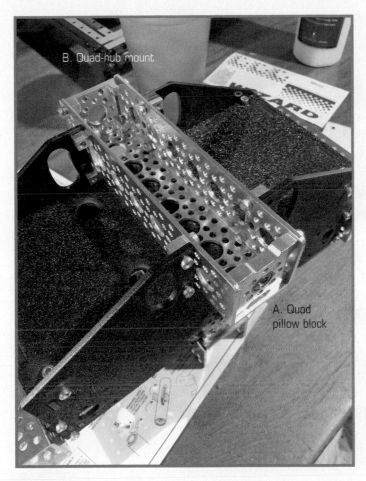

B. Quad-hub mount

A. Quad
pillow block

FIGURE 13.12 This 6" channel will eventually support the wheel assemblies.

4. Next, grab the twin 12" channels and install the four motor mounts to either end, using more of those handy quad-hub mounts. Figure 13.13 shows a couple of mounts already installed.

FIGURE 13.13 Attach the motor mounts to the 12" channels.

5. Install the set screw hub and 8" axle to one 12" channel, as shown in Figure 13.14.

6. Attach one 12" channel and secure it to the quad-hub mount on the 6" channel mounted to the ABS chassis. Use a spacer and bearing on the 12" channel where the shaft passes through. You can see how it should look in Figure 13.15. This channel will want to move around, and that's great! It will help the rover traverse rough terrain.

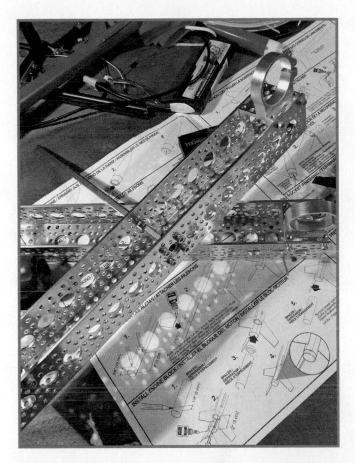

FIGURE 13.14 Attach a set screw hub and an 8" axle to one of the 12" channels.

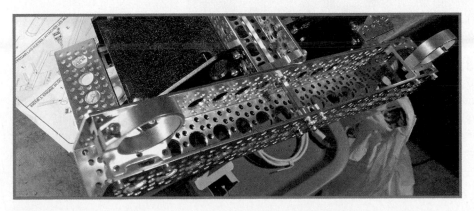

FIGURE 13.15 Connect one of the 12" channels to the chassis.

7. Attach the other 12" channel, giving it its own bearing to support the other end of the axle, as shown in Figure 13.16. This channel doesn't move, which might alarm people, but it makes sense—if both wheel-mounts moved, the chassis would fall to one side or the other.

FIGURE 13.16   Attach the other 12" channel.

8. Install the motors and hex wheel adapters, as shown in Figure 13.17. The motors simply slide into the clamp mounts on the ends of the 12" channels, which can then be tightened to secure them. Onto each axle add a hex wheel adapter, onto which the wheels will be attached.

FIGURE 13.17 Add the motors and wheel adapters.

9. Assemble the wheels and then install them. They consist of a plastic rim, a foam insert that fits around the rim, and a rubber tire that covers the foam. You can see how they should look in Figure 13.18. Use more of the #6-32 socket-head screws to attach the wheels to the adapters.

FIGURE 13.18 Assemble the wheels, then install them.

10. You're done with the kit! It should now look like Figure 13.19. In the next step you'll customize the rover by adding the sensors. Note that all the Acrobatics parts you'll add after this step have to be purchased separately, as described in the parts list.

11. Connect the servo to the servo mount using #6-32 screws (see Figure 13.20).

FIGURE 13.19   You're done with the kit! Now to customize it.

FIGURE 13.20   Attach the servo to the servo mount.

12. Install the servo in the top channel, as shown in Figure 13.21. Keep the servo's hub as close to the middle of the channel as you can. The servo's wires can be stuck through the center hole of the channel.

FIGURE 13.21 Install the servo in the top channel.

13. Install the servo hub using the set screw that came with the servo (see Figure 13.22).

FIGURE 13.22 Install the servo hub.

14. Attach a 3" length of channel to the servo hub, as shown in Figure 13.23. This will be the turret that rotates the ultrasonic sensor.

FIGURE 13.23 Add a 3" length of channel.

15. Assemble the ultrasonic sensor's mount. This advanced system, shown in Figure 13.24, consists of a piece of wood with the sensor attached to it with double-sided tape. While you're at it, attach wires to the sensor's four leads. The wood is 1.5" by 3", thus allowing it to fit inside the channel once installed.

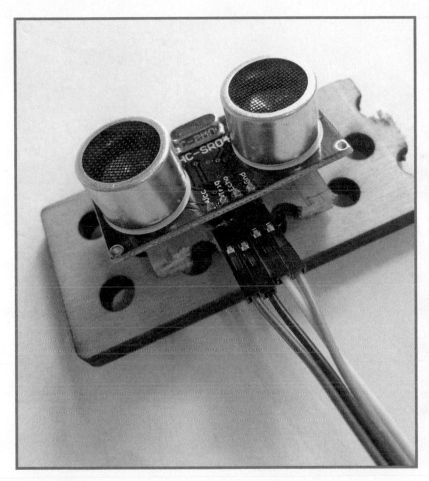

FIGURE 13.24 Assemble the ultrasonic sensor's mount.

16. Install the ultrasonic sensor assembly in the 3" channel. Use #3 pan-head screws to secure the wood, or else just use the friction method: The pieces of wood I used were exactly the right width to be held in place without hardware. Figure 13.25 shows how it should look.

FIGURE 13.25 Install the ultrasonic sensor assembly.

17. Next, you'll need to wire up the RFID sensor. One optional tool that might help is an RFID breakout board, the one I mentioned in the parts list. This breaks out the sensors' 2mm-spaced pins into a breadboard-friendly 0.1". Either way, Figure 13.26 shows which pins get wires; later on in the build I'll show you where to attach the other ends of those wires.

FIGURE 13.26   Wire up the RFID module as you see here.

18. Stick the sensor to a piece of wood as you did with the other assembly. Figure 13.27 shows how it should look.

19. Next, let's begin installing the electronics. Figure 13.28 shows the various components and how to connect them, but let's cover each item on its own:

   a. Seat the motor shield on the Arduino.

   b. Attach the motor wires to the four motor ports on the Motor Shield. These simply connect to screw terminals. If a motor's rotation is opposite what you're expecting, simply reverse the leads.

   c. Connect the servo's wires to the triple pins on the Motor Shield. The plug on the end of the wires should push down onto the pins.

   d. Wire up the ultrasonic sensor. The pins marked 5V and GND are attached to their respective pins on the Arduino. TRIG connects to pin 12, and ECHO connects to pin 11.

FIGURE 13.27 Attach the sensor to the rover.

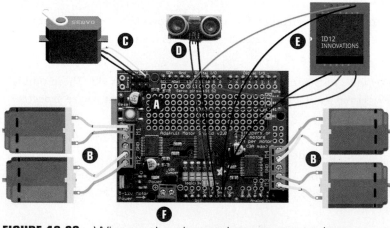

FIGURE 13.28 Wire up the electronics as you see here.

e. Connect the RFID's wires to their respective pins on the shield. Refer to Figure 13.26 to see how to wire up this component.

f. Attach the batteries. The 9V battery and its plug connect to the barrel jack on the Arduino, while the 12V battery pack plugs into the screw terminal on the shield marked 5-12V Motor Power.

Programming the RFID-Navigating Rover

The Rover's sketch includes a number of complicated components: It can read RFID tags, of course, and it's constantly scanning for tags and then comparing captured codes with a set of actions it can undertake.

For instance, if you scan one code, the Rover turns the servo that controls the direction in which the sensor is aimed, and then performs a distance scan. Another tag sends the Rover backwards, whereas another rotates it in place. These are all stand-ins for your own directions, which will be unique to the environment you wish to navigate. The purpose of these examples is simply to show you how to create your own.

```
//This code was created with example code from Adafruit's Motor Shield library
//as well as Bildr.org's RFID tutorial
#include <Servo.h>
#include <AFMotor.h> //you may need to download the Adafruit Motor Shield library
int RFIDResetPin = 13;
//this is the ultrasonic's two control pins.
//if you use a 3-pin PING, simply change their Arduino pin number
//so that they're the same.
int us_ping - 12;
int us_listen = 11;
// initializes the four motors. FL - front left, and so on.
AF_DCMotor FLmotor(1);
AF_DCMotor FRmotor(2);
AF_DCMotor BLmotor(3);
AF_DCMotor BRmotor(4);
//initialize the servo
Servo USservo;
//give each tag its own array — obviously, add the correct codes
//the 13 refers to the size of the array, 12 characters plus a null
char tag1[13] = "000000000000";
char tag2[13] = "111111111111";
char tag3[13] = "222222222222";
void setup() {
  Serial.begin(9600);
  pinMode(RFIDResetPin, OUTPUT);
  digitalWrite(RFIDResetPin, HIGH);
```

```
   //power up the servo and four DC motors
   USservo.attach(10); //AKA servo 1 on the motor shield
   FLmotor.setSpeed(200);
   FLmotor.run(RELEASE);
   FRmotor.setSpeed(200);
   FRmotor.run(RELEASE);
   BLmotor.setSpeed(200);
   BLmotor.run(RELEASE);
   BRmotor.setSpeed(200);
   BRmotor.run(RELEASE);
   //initialize the US sensor's pin
   const int us_Listen = 11;
   const int us_Ping = 12;
   long duration, inches, cm;
}
void loop() {
   char tagString[13];
   int index = 0;
   boolean reading = false;
   while (Serial.available()) {
     int readByte = Serial.read(); //read next available byte
     if (readByte == 2) reading = true; //begiNning of tag
     if (readByte == 3) reading = false; //end of tag
     if (reading && readByte != 2 && readByte != 10 && readByte != 13) {
       //store the tag
       tagString[index] = readByte;
       index ++;
     }
   }
   checkTag(tagString); //Check if it is a match
   clearTag(tagString); //Clear the char of all value
   resetReader(); //Reset the RFID reader
}


switch (tagString) {
case 'tag1':
   servo1.write(90); //rotates the sensor
   //sends out a ping
   pinMode(us_Ping, OUTPUT);
   digitalWrite(us_Ping, LOW);
   delayMicroseconds(2);
   digitalWrite(us_Ping, HIGH);
   delayMicroseconds(5);
```

```
  digitalWrite(us_Ping, LOW);
  //then listens
  pinMode(us_Listen, INPUT);
  duration = pulseIn(us_Listen, HIGH);
  cm2 = microsecondsToCentimeters(duration);
  if (cm2 < 1000) //triggers within 1 meter
  {
    //ADD SOME EVENT THAT OCCURS IF SOMETHING'S WITHIN 1M
  }
  servo1.write(-90); //returns to its original position
  break;
case 'tag2':
  //MOTORS MOVE BACKWARD FOR 2 SECONDS
  FLmotor.run(BACKWARD);
  FLmotor.setSpeed(200);
  FRmotor.run(BACKWARD);
  FRmotor.setSpeed(200);
  BRmotor.run(BACKWARD);
  BRmotor.setSpeed(200);
  BLmotor.run(BACKWARD);
  BLmotor.setSpeed(200);
  delay(2000);
  FLmotor.setSpeed(0);
  FRmotor.setSpeed(0);
  BRmotor.setSpeed(0);
  BLmotor.setSpeed(0);
  delay(5);
  break;
case "tag3":
  //turn in place
  FLmotor.run(FORWARD);
  FLmotor.setSpeed(200);
  FRmotor.run(BACKWARD);
  FRmotor.setSpeed(200);
  BRmotor.run(BACKWARD);
  BRmotor.setSpeed(200);
  BLmotor.run(FORWARD);
  BLmotor.setSpeed(200);
  delay(2000);
  FLmotor.setSpeed(0);
  FRmotor.setSpeed(0);
  BRmotor.setSpeed(0);
  BLmotor.setSpeed(0);
```

```
    break;
  }
clearTag(tagString); //Clear the char of all value
resetReader(); //reset the RFID reader
}
//this function converts microseconds to centimeters
long microsecondsToCentimeters(long microseconds)
{
  return microseconds / 29 / 2;
}
//this resets the reader
void resetReader() {
  digitalWrite(RFIDResetPin, LOW);
  digitalWrite(RFIDResetPin, HIGH);
  delay(150);
}
// this function clears the tag's code by filling the array with zeros
void clearTag(char one[]) {
  for (int i = 0; i < strlen(one); i++) {
    one[i] = 0;
  }
}
```

Summary

You finished your final drone project in this chapter, building a rover out of an Actobotics Nomad chassis with an Arduino running the show. The main quadcopter project of the book isn't complete, however: You must still program it. Chapter 14, "Building a Quadcopter VI: Software," introduces you to some popular drone-control programs that allow you to control your craft from a laptop or tablet. After that, I'll walk you through the process of configuring the MultiWii flight controller mounted on your quadcopter. Soon thereafter you'll be flying!

14

Building a Quadcopter VI: Software

You've reached the final chapter of the book, in which you learn about software you can use to control your quadcopter. This chapter covers a variety of flight control software suites, including applications for your laptop and your favorite mobile devices. Think of these as autopilots that enable you to control your craft without the usual joysticks. After that, you'll configure your MultiWii, getting it ready to fly (see Figure 14.1).

FIGURE 14.1 The quadcopter project reaches its logical conclusion: in the air!

Flight Control Software

As you might expect, software can control a motor even better than your thumb can. Leaving the control of your drone to software enables you to operate the camera or monitor the GPS and other sensors. In some of these packages, you can plan out a trip so that the drone follows a set flight pattern.

One thing to keep in mind is that typical flight control packages are configured for a single type of flight control board. It makes sense if you think about it—how could a piece of software know about the specific configuration of your drone? In any case, manufacturers quite naturally want you to use their software to control their drones.

OpenPilot

This open-source, nonprofit platform focuses on providing flight stabilization and autopilot software for multicopters and any other types of autonomous vehicles, including tricopters, fixed-wing planes, and rovers. Each type of craft is supported with its own installation instructions, thus allowing a Y-configuration hexacopter (for instance) to be controlled as easily as a more commonplace X-configuration, the same kind as the drone you'll build in this book.

As an open-source project, OpenPilot was created by an entire community of developers and made its debut in 2010 (see Figure 14.2). Since then, the community has developed different hardware platforms to go along with the software, as well as adding on robust features to the autopilot itself. To learn more about OpenPilot, visit openpilot.org.

FIGURE 14.2 OpenPilot.

MultiWii

Like OpenPilot, MultiWii was developed by an open-source community (see Figure 14.3). The original project involved using a Wii nunchuck as a controller (hence the name), but it now has a solid multicopter platform with hardware and software features that are added with each version released.

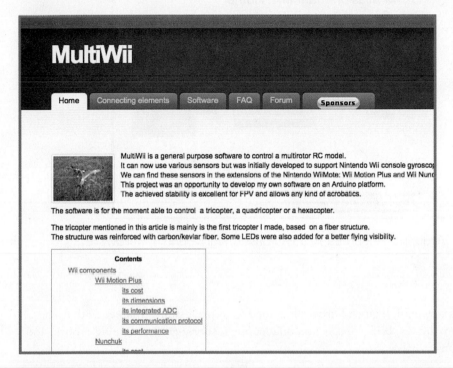

FIGURE 14.3 The MultiWii site shows how to control a drone with a Wii nunchuck.

At first, the hardware consisted of off-the shelf modules such as Arduinos and IMU boards, but an official flight controller board was finally developed. You can buy a clone of the project from a variety of online stores, or you can piece together the parts yourself. To learn more about MultiWii, download software, or to find out how to participate, visit multiwii.com.

APM Planner 2.0

In the heady days of DIY drones, the first attempts to build Arduino-based autopilots were called Ardupilot. As time passed, the software was incorporated into products sold by drone manufacturer 3DRobotics, and the overarching software project was renamed APM.

APM, shown in Figure 14.4, supports fixed-wing planes, rovers, and a bevy of copters featuring between two and eight propellers in standard configurations. You can learn more at planner2.ardupilot.com/

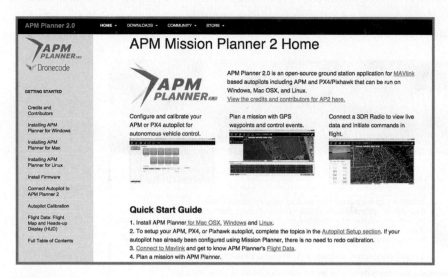

FIGURE 14.4 Ardupilot began as Arduino-based flight-control hardware.

eMotion

SenseFly offers powerful, high-end drones intended for a professional market, with fixed-wing agriculture drones, bridge-inspection quadcopters, survey-grade mapping drones, and other non-hobbyist hardware.

Not surprisingly, their eMotion software, shown in Figure 14.5, is serious business, with high-end features such as flight simulation, mission planning, and sensor monitoring through the app. The downside is that eMotion only works with SenseFly's drones! SenseFly's website, sensefly.com, has information on how to download eMotion.

AR.Freeflight

Parrot has long been at the forefront of hobbyist drones, and their AR Drones, featuring a distinctive cloverleaf design, are some of the most popular commercial drones in the world. Parrot's drone-control app AR.Freeflight, shown in Figure 14.6, controls all their drones, enabling pinpoint control with a phone application.

With a nod to the phenomenon of the drone as a modern-day photography platform, AR.Freeflight has an optional add-on called Director Mode that optimizes the flying experience for moviemaking, with built-in motions mimicking the behavior of panning, crane shots, and flight stabilization. You can learn more at parrot.com.

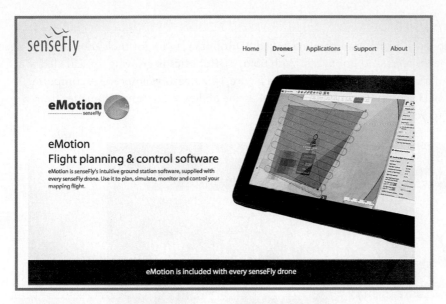

FIGURE 14.5 SenseFly drones are intended for serious tasks, and their software, eMotion, is just as robust.

FIGURE 14.6 AR.Freeflight serves as the official control system for Parrot drones.

3DR Solo App

Earlier, I mentioned a drone manufacturer called 3DRobotics, who got their start by catering to DIY drone builders and hobbyists. With Solo, 3DRobotics is growing up. Labeled the "world's first smart drone," the Solo, shown in Figure 14.7, features impressive computing power (with a 1GHz computer onboard) as well as the ability to stream HD images from a GoPro camera to your favorite mobile device.

FIGURE 14.7   The 3DR Solo app controls the entire Solo experience.

When they say it's "smart," they're not kidding around. Follow Me mode makes the drone follow a single target around. The app is smart, and can automatically generate a repair ticket when the drone crashes. You can find out more about all of their offerings at 3DRobotics.com.

Configuring the MultiWii

Let's configure our flight controller of choice, the MultiWii. Once again, you can learn all about the project at Multiwii.com. Let's get started!

1. Ensure that you have the latest and greatest version of the Arduino IDE downloaded and installed.

2. Find the MultiWii software from the project's code repository at https://code.google.com/p/multiwii/ and download it.

3. Open the MultiWii sketch in the Arduino environment. The sketch consists of several tabs, as discussed later in the chapter. Click the tab marked "Config.h" and customize the sketch based on your copter configuration, RC system choice, and microcontroller

selection. The instructions in config.h are straightforward, and you should have no trouble customizing your setup.

4. Connect the MultiWii to your PC, using a micro-USB cable, as shown in Figure 14.8.

5. Upload the sketch to the MultiWii board using the normal Arduino rules (the MultiWii is essentially just a customized Arduino board.)

6. Configure your transmitter. This mostly just means setting the maximums of the usual four controls: throttle, pitch, roll, and yaw.

7. Tune the PID. This involves fine-tuning the controls, and you can learn how to do it here: http://www.multiwii.com/wiki/index.php?title=PID.

8. Calibrate the sensors. This requires interfacing your computer with the FC using a software interface such as WinGUI (https://code.google.com/p/mw-wingui/). The GUI allows you to calibrate the magnetometer, accelerometer, and gyroscope.

9. Conduct a test flight, as described in the final section of this chapter.

FIGURE 14.8 Use a USB cable to connect the MultiWii to your computer.

You're done! In the next section, you'll configure the actual drone for a test flight.

Examining the MultiWii Control Sketch

Although I'm not going to delve too deeply into each portion of the sketch, I would like to give you an overview of the sketch's architecture. This will help you with troubleshooting or customization.

Open the file named multiwii.ino in your Arduino IDE. You should see a simple welcome message as well as a number of tabs containing the actual components of the sketch (see Figure 14.9). Let's go through these sub-sketches one by one:

- **Alarms.cpp and Alarms.h**—This library controls the buzzer and various alert LEDs on the MultiWii. As you might recall, libraries in the Arduino world consist of a source file (.cpp) and a header file (marked with an .h), and they provide critical functions.
- **EEPROM.cpp and EEPROM.h**—This library manages the storage GPS waypoints in the MultiWii's memory.
- **GPS.cpp and GPS.h**—As you might expect, this sketch controls the multicopter's GPS capability.
- **IMU.cpp and IMU.h**—This library manages the inertial measurement unit (IMU). This is the sensor that helps the drone determine its compass heading and altitude.
- **LCD.cpp and LCD.h**—Some quadcopter configurations allow the pilot to interface with the MultiWii with the help of a LCD screen, and this library manages the LCD.
- **MultiWii.cpp and MultiWii.h**—This library contains the core functions of the drone, pulling in data and functions from the various other libraries that make up MultiWii.
- **Output.cpp and Output.h**—This library controls the multicopter's motors and servos, and the settings for each possible configuration can be found here.
- **Protocol.cpp and Protocol.h**—MultiWii uses MSP (MultiWii Serial Protocol) to communicate with its various components, and the Protocol library governs MSP.
- **RX.cpp and RX.h**—This library is another resource that supports serial communication.
- **Sensors.cpp and Sensors.h**—This library manages sensor input, including the accelerometer, magnetometer, gyroscope, and barometer, among others.
- **Serial.cpp and Serial.h**—This library is the main serial control resource.
- **Config.h**—You'll fine-tune the copter's settings in this library, including selecting the copter type, using an alternative CPU, and setting up your radio control settings.
- **Def.h**—This library is chock full of definitions: named constants that work in the background of the sketch.
- **Types**—Another library of cryptic code that helps the MultiWii do its thing.

FIGURE 14.9 The MultiWii's Arduino sketch includes a bunch of individual files organized in tabs.

Pre-Flight Checklist

You're almost ready to test the quadcopter! Go through this simple pretest checklist before you begin:

1. When you arrive at the flight area, check out the vicinity for power lines, big trees, and other drone-killing obstructions. You should also be aware of the proximity of restricted airspace such as airports.

2. Find a flat, open area to launch and land the drone.

3. Connect a charged battery to the drone's wiring harness. While you're at it, double-check the various wires to make sure nothing's loose.

4. Power on the craft while keeping the throttle to zero. Many a quadcopter has leaped off the landing pad at full velocity because the throttle was maxed out.

5. Check the props and verify rotational direction. The copter won't fly right if the props aren't rotating the direction they're supposed to.

6. If you have a camera mounted on the drone, power it up. If you're going to be shooting video, begin recording.

7. Launch! Hit the throttle slowly and increase speed judiciously, getting a sense of the way the drone handles as it rises off the ground, as shown in Figure 14.10.

FIGURE 14.10 The quadcopter hovers a few feet off the ground.

Summary

With the software in place, you're ready to hit the skies! I hope this book was useful in your explorations of drone technology. Good luck and good flying!

Glossary

3D printer A machine able to extrude and deposit layers of plastic in order to form a three-dimensional object.

Accelerometer A sensor that determines its speed and acceleration and returns that value to a microcontroller.

Airframe The chassis of a quadcopter.

Amp The unit of electrical current.

Analog Data sent in a continuous wave of varying voltage, as opposed to digital, which sends data with a series of on-and-off signals.

Array In programming terminology, an array is a list of values stored for future use.

Autonomous robot A robot that relies on its own programming, rather than a human operator, to make control decisions.

Autopilot The microcontroller that steers a drone along a pre-programmed flight path.

Barometric sensor A sensor that detects changes in air pressure, much the way a barometer does.

Board A shorter way of saying a *printed circuit board* (PCB).

Breadboard A hole-punched plastic board with concealed conductors, allowing you to wire up circuits easily and without solder.

Breakout board A small printed circuit board (PCB) used for controlling a single component. For instance, you can create a breakout board for managing an L293D motor control chip.

Brushed/brushless motor A distinction between the types of DC motors, with the difference being how electricity is transmitted to the motor's windings.

Bushing A small fastener used to secure an axle.

Carbon-fiber A strong and lightweight material often used in drones and other aircraft.

Chassis The frame onto which the components of a robot or drone are affixed.

Compile To convert one computer language to another, typically used to turn people-readable code to machine-readable code.

Computer numerically controlled (CNC) tools Rail-mounted power tools that precisely follow paths as directed by a computer program.

Datalogger   A module in a microcontroller project that records data—for instance, a sensor's readings.

DC motor   A commonplace motor that rotates its hub when voltage is applied to its terminals.

Digital   A type of data that consists exclusively of yes-or-no instructions, versus analog data, which consists of varying voltage levels.

Electronic speed controller (ESC)   A device that triggers high voltage in a motor, responding to a low-voltage signal from a microcontroller or radio receiver.

Encoder   A device that can detect how far a motor's hub has turned, and returns this value to a microcontroller.

First-person video (FPV)   Live video, typically low resolution, that allow a drone's pilot to see what the drone sees.

Flight controller   A microcontroller board optimized for controlling a drone with the help of altimeters, magnetometers, and other sensors.

Ground   The return path of an electric circuit. On a battery, the ground is marked with a – (minus sign). Ground is often abbreviated GND in electronic parlance.

Ground bus   The strip of conductor on a breadboard that is usually marked as black or blue and designated as the ground.

Heat shrink tubing   Non-conductive rubber tubing used to cover wire joins. As heat is applied, the tubing shrinks down to cover up the exposed wire.

Infrared (IR) light   A bandwidth of light outside of the visible range for humans, IR light is often modulated to send small amounts of data—for instance, the "off" signal for a TV.

Initialize   To create a new variable and assign it a value.

Inrunner   A motor featuring a rotating axle surrounded by fixed electromagnets.

Integrated development environment (IDE)   Software that provides technical services to programmers to assist them in creating code.

Integrated circuits (ICs)   A series of circuits miniaturized and then embedded in a plastic housing.

IR receiver   Sensor that detects infrared light pulsed at the correct frequency, 38MHz.

Jumper   A generic term for wires or conductors used in electronics projects.

Laser cutter   Also known as a *laser etcher*, a laser cutter burns through thin materials such as cardboard, medium-density fiberboard (MDF), and particle board.

Lead   A wire or terminal on a component to which a wire is attached.

Light-emitting diode (LED)   The LED is the light bulb of the electronics world.

Library   Supporting code referenced by an Arduino sketch, allowing you to keep the main sketch relatively simple.

Light sensor A sensor that detects light. Some of these operate as a variable resistor, where the level of light dictates resistance, whereas others are digital and send numeric data to the microcontroller.

LiPo A lithium polymer battery, a type of rechargeable battery used in robotics.

Magnetometer A sensor that detects magnetic fields, particularly the Earth's.

Mechanum wheels Wheels with smaller wheels along the rim, allowing a robot to move sidewise as well as forward and backward.

Mesh network A network consisting of multiple nodes, each able to see every other node.

Microcomputers Miniature computers, with all the capabilities of full-sized PCs, if not the specs.

Microcontroller A simplistic computer, able to take input from sensors and activate motors and lights.

Motor control chip An integrated circuit optimized to control motors, expanding on the Arduino's capabilities.

Multicopter Generically, a quadcopter or a copter with more or fewer props, such as a tri-copter or octocopter.

NiCad Nickel–cadmium, a type of rechargeable battery.

NiMH Nickel–metal hydride, a type of rechargeable battery.

Omni wheel A drive wheel studded with free-rotating side wheels, allowing the main wheel to roll, unpowered, perpendicular to its powered direction. Also known as a *mechanum wheel*.

Open-source hardware and software Electronics projects where the code and electronic designs are shared freely, and anyone is free to modify or re-create them.

Outrunner A motor whose outer casing and electromagnets rotate around a central axle.

Passive infrared (PIR) sensor An infrared sensor that detects movements via subtle changes in temperature.

Pin The power and data connectors of an Arduino.

Potentiometer Usually referred to as a *pot*, a potentiometer is variable resistor that's adjusted via a knob.

Power bus The conductor strip on a breadboard designated to supply voltage to the board.

Printed circuit board (PCB) A composite board coated in a conductive material, enabling you to etch circuits onto the board and thereby create electronic assemblies.

Prop saver A style of breakaway propeller mount that helps save the prop from damage.

Pulse-width modulation (PWM) A way of "dimming" an electronic component such as a motor or LED that is normally either "on" or "off" by rapidly flickering it on and off.

Quadcopter A small aircraft consisting of four propeller-mounted motors, usually quadri-laterally mounted.

Radio control (RC) system   A control system for a robot or model vehicle, consisting of a controller and receiver.

Real-time clock (RTC) module   A timekeeping chip with a battery backup, designed to maintain the correct time for several months.

Remote operating vehicle (ROV)   A tethered underwater drone used for undersea exploration.

Resistor   An electronic component designed to limit the flow of electricity to protect fragile components and control the flow of voltage in the circuit.

Radio-frequency identification (RFID)   A system that includes a powered sensor able to read unpowered tags.

Rotary tool   A small power tool with multiple types of attachments, ranging from saws to sanders to polishers. You've probably heard of the category leader, Dremel.

Rover   A ground-traversing drone in the shape of a car.

Schematic   The drawn representation of a circuit, with symbols representing the various components.

Sensor   An electronic device that sends data or voltage to a microcontroller about the environment around it.

Serial communication   A method of communication whereby data is sent along a single wire, with each bit sent sequentially.

Serial monitor   The window in the Arduino integrated development environment (IDE) where serial traffic can be monitored. This can be a great tool for debugging programs.

Servo   A motor equipped with a gearbox and encoder, enabling precision control of how far the motor's shaft turns.

Servo horn   The disk or lever that attaches to the servo's rotor, as well as the thing the servo is to move.

Shield   An add-on circuit board for the Arduino. It stacks right on top, sharing the Arduino's pins while adding additional capabilities.

Sketch   Arduino parlance for the program that controls the Arduino's pins.

Standoffs   Metal or plastic inserts that are often used to create space or support between a printed circuit board (PCB) and another surface.

Stepper motor   A motor designed to rotate in increments, called *steps*. It usually has four or more wires.

Temperature and humidity sensor   A digital sensor that measures temperature and humidity and returns a numeric reading to the microcontroller.

Terminal strips   The rows of connectors in breadboards, running perpendicular to the power and ground bus.

Toolpath   The path followed by a computer-controlled tool.

Transistor   A miniature electronic switch controlled with electrical signals.

Ultrasonic sensor   A sensor that detects obstructions and measures distances by transmitting a beam of inaudible sound and then listening for an echo.

Unmanned aerial vehicle (UAV)   The proper name for an aerial drone.

XBee   A wireless module using the popular ZigBee protocol, which is often used for home automation.

Index

Symbols

A

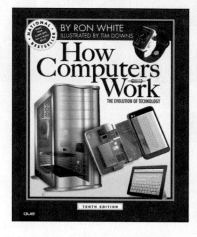